# INSTRUCTOR'S MANUAL

### AND

## TEST BANK

to accompany

# ELEMENTS OF ECOLOGY
### *Third Edition*

Robert Leo Smith
*West Virginia University*

and

Edmund E. Bedecarrax
*Department of Biology*
*City College of San Francisco*

HarperCollinsCollegePublishers

Instructor's Manual and Test Bank to accompany ELEMENTS OF ECOLOGY, Third Edition

**Copyright © 1992 by HarperCollins College Publishers**

ISBN: 0-06-366334-1

92 93 94 95 96  9 8 7 6 5 4 3 2 1

# CONTENTS

# PREFACE

Some textbooks come with an instructor's manual; some do not. Ecology texts generally belong to the latter group. The usefulness of such manuals depends upon the amount of effort and thought put into them. I have tried to make this manual a useful one, to be a guide to the text. The introduction lays out the plan of the text and offers suggestions on how to use the text. Overall, the text may be too long to be completed in any one semester. But to write a shorter one would destroy any flexibility for its use in the classroom. Each one of us has our own approach to teaching ecology. What one instructor may emphasize, another will not. By providing sufficient material, the textbook can be molded by the instructor to fit the course. I offer suggestions how to use the flexibility built into the text.

For each chapter I provide a commentary which stresses some of its essential points, where appropriate how the material relates to topics covered in other chapters. The commentaries also offer some kernels of ideas to include in lectures. Following the commentary is a section on Source Materials giving references, both articles and books, in addition to those listed in Selected References at the end of each chapter. The Discussion Topics supplement the Review and Study questions at the end of each chapter. These topics are strongly applied and are aimed at tying ecology to the economic, political, and behavioral aspects of environmental problems. Given the diversity of students taking environmentally-oriented ecology courses, these topics should appeal to students in other disciplines such as history, economics, literature, psychology, and the like. Some topics will require thought and library research.

The bulk of the manual consist of a test bank of over 1000 comprehensive questions developed by Dr.Edmund E. Bedecarrax of Department of Biology, City College of San Francisco. These questions, which include sentence completion, multiple choice, true and false, matching and discussion will be a godsend to harried instructors. With this huge number of test questions from which to draw, instructors will be able to vary their tests from semester to semester. Could instructors ask for more?

Robert Leo Smith

# INTRODUCTION

# INTRODUCTION

One problem facing an instructor who plans to adopt a new text is how best to to use the book. Will I have to rewrite my lectures to fit the text , or reorganize my course? How you answer that question depends a good deal on your own approach to teaching and the role a textbook plays in your course.

There are roughly three approaches. One is to base the course on the textbook and to follow its contents chapter by chapter to the end. A second approach is to use the textbook as a source of information that supplements the lecture material. Between the two, the student is exposed to a maximum amount of subject content. A third approach is to require the student to buy a textbook and after the first day of class ignore it. Students then depend wholly upon the lecture material for the course. In the later approach, it would be better not to require the student to own a text. It makes for better relations among students, instructors, and book stores.

The first approach suggests the instructor must include all of the textual material in the course. If that material is more than can be covered in a semester, students and instructor conclude that the text is too long.

The second approach places the textbook in its proper perspective. The textbook should supplement the lecture by providing the additional examples and explanations needed to expand the lecture; or the textbook can serve as a focal point on which additional material and examples can be added in the lecture. Between the textbook and the lecture, as well as the laboratory, if one is offered, the student should receive a good basic understanding of the subject.

For the instructor, the textbook should provide a survey of ecology, summarizing concepts and theories and providing examples of and ingress into the literature of unfamiliar areas of ecology. The literature of ecology has expanded so greatly that few instructors have the time or the motivation to explore areas of ecology outside of their own research interests.

ELEMENTS OF ECOLOGY is relatively a short text. In spite of its nearly the same number of pages as the Second Edition, the Third Edition is 20 percent shorter in content matter, but retains a balance in subject matter. Yet for some the text may still seem to be too long for a semester. This is deliberate on my part. The major problem facing a textbook author is what to include and what to omit. If you omit certain topics, the book will appear incomplete or idiosyncratic. A very short text gives the impression that the author dictates what should be included in the course and the instructor may have to fall back on other texts or sources for material omitted. If you cover too much, the discussion becomes shallow. Because no one teaches ecology in exactly the same manner. Because of the wide range of subject matter in ecology, an one instructor may stress certain topics and omit others, depending upon the emphasis of the course. Your inclusions and omissions will not be the same as those of another. My aim is provide the diversity and flexibility needed to cover a range of approaches to ecology.

## Plan of the Text

ELEMENTS OF ECOLOGY, Third Edition, has been completely revised and rewritten. Its focus is strongly applied with an environmental emphasis. It is intended mostly for a nonmajors ecology course with an applied environmental emphasis or an environmental science course with strong ecological approach. Thus this text fills a gap in general ecology texts, most of which are weak in applied ecology or ignore it altogether and are aimed toward students in disciplines other than biology. Their main needs are a general appreciation of ecological principles and application to real world problems. Environmental science texts are strong on environmental issues but their approach is heavily sociological and political. They are weak in discussing ecological principles and fail to explore the ecological concepts and principles involved. ELEMENTS OF ECOLOGY provides a solid introduction to ecological principles and their application to environmental issues. These applications are common thread throughout the text and are not isolated in one or two separate chapters. This approach is unique to ELEMENTS OF ECOLOGY.

ELEMENTS OF ECOLOGY features several pedagogical innovations for an ecology text. Each chapter opens with an outline and a list of objectives for the student. Each chapter ends with a summary, a list of review and study questions, and suggested readings, books the reader may consult for further study. The purpose of the review and study questions is two-fold. Study questions are designed to encourage the student to master the subject matter in each chapter. The most effective use of the study questions results when the students answer the questions for themselves ahead of the lectures on the topic. If you require the students to answer the study questions, they will obtain much of the course material from their own reading. That leaves you free to lecture on certain points or concepts you wish to emphasize or to engage the class in discussions. Some questions are starred (*). These questions are intended to encourage students to apply ecological concepts studied to current environmental problems.

ELEMENTS OF ECOLOGY is divided into five parts, each building upon the other. Part I consists of a lone chapter. It defines ecology, discusses briefly its diverse origins, examines ecology as a science and the application of ecological principles to real world problems.

Part II introduces natural selection, population genetics, and speciation. Students, and perhaps some instructors, may wonder about the reason for introducing these topics so early in the text or why they should be included at all. I have two reasons for their early placement. Appreciation of natural selection is basic to understanding the adaptations of organisms to environment discussed in Part III. Natural selection involves population genetics, so most logical place to introduce that topic, essential to an elementary understanding of such current environmental controversies as spotted owl, Florida panthers, decline in neotropical warblers, among many others. Much of the debate about species preservation and biodiversity centers about the nature of species and differences among them. All students should have a understanding of the species concept early.

Part III is ecology at the organismal level, often termed autecology. It is mostly physiological ecology with emphasis on the adaptive responses of organisms to temperature, moisture, and light. The Introduction presents the concepts of the law of the minimum and law of tolerance, homeostasis, and systems. Chapter 4 looks at the radiant solar energy and its effects on global, local, and micro- climates. Following that are chapters on adaptations of organisms to variable thermal and moisture environments. Light is considered from two aspects: its role in photosynthesis and physiological responses of plants to light and in daily and seasonal periodicity of plant and animal activity. Part III concludes with a consideration of the nature and properties of soil, the foundation of terrestrial life on Earth.

Part IV is devoted to population ecology. Its objective is to enable students to understand the nature of population growth, regulation, and interactions, as they apply to real world problems from explosive growth of pest populations and extinction of species to the spread of disease and overexploitation of natural populations. The Introduction emphasizes the differences between plant and animal population which should be appreciated before moving into the main material. The initial chapters covering distribution, age structure, mortality, natality, and growth are short to emphasize various population characteristics and to break up some of the more difficult material. These topics are followed by population regulation and intraspecific competition, reproductive effort, r and K selection, and population interactions including interspecific competition and predation in its broadest sense, and the role of predation in the population dynamics of a species. Chapter 17 considers the place of two symbiotic relationships, parasitism and mutualism, in interspecific population dynamics. Part IV ends with Chapter 18 on the impact of human exploitation of natural populations and the extinction of species.

Part V advances to the Community. This section covers the concept of the community, spatial and temporal patterns of community structure, species diversity, island biogeography, the role of disturbance in community change and stability, and the effects of human impacts on community development.

Part VI looks at Ecosystem Dynamics. The Introduction calls attention to the ecosystem concept and is basic to the chapters that follow, covering energy flows and production in the ecosystem, food webs and the role of the various functional groups, including decomposers, and nutrient cycling. The last chapter, 25, explores major human intrusions in ecological cycles.

Part VI is a survey of ecosystems: terrestrial, aquatic, and marine, their structure, function, and associated environmental pressures and problems. Some ecologists prefer to ignore any discussions of ecosystems per se, arguing that it is largely descriptive ecology and that basic concepts and problems are general enough to be applicable to any ecosystem. Although basic ecological principles hold for various types of ecosystems, their applications differ. Consider, for example, nutrient cycling. The basic principles of nutrient cycling hold true for all ecosystems, but the manner in which they are cycled are quite different. There are functional differences between aquatic and terrestrial ecosystems; and there are major functional differences between flowing water ecosystems and still water ecosystems. These differences, both structural and functional, are reviewed in this last section of the book.

This final part should form an important part of any nonmajors course. Many of our current environmental controversies center about specific ecosystems: logging of old growth forest, tropical deforestation, wetland drainage, damming of rivers, channelization of streams. The utilitarian forces causing the problems have absolutely no knowledge of the ecology of the systems they are destroying. The environmentalists lack ecological expertise to come up with sound ecological arguments, so they resort to emotionalism.

## How to Use the Text

I revised ELEMENTS OF ECOLOGY with one question in mind: With what ecological concepts should nonecologist--doctors, lawyers, engineers, accountants, business managers, journalists, and liberal arts majors--be familiar to understand today's environmental problems? The contents of the text reflect my answer to that question. The answers is based on 26 years of experience of writing a weekly regional natural history and ecology feature column and reader responses to them. Some users may argue that the amount of material is too much to cover in one course; it probably is. But the material presented is sufficient for instructors to tailor the text to fit their own particular way of covering the course. For a strong environmentally oriented course, I would recommend the inclusion of Chapters 1, 2, 3, 4, 9, 10, 11, 12, 13, 16, 17, 18, 19, 20, 21, 22, 23, 24, 25, 26 though 36, with references to selected topics in other chapters, especially 5, 6, and 7. For a population oriented course the following chapters should be included 1, 2, 7, 8, 10 through 21, and selected topics from Chapters 25 through 36. To emphasize organismal responses to environmental change instructors could use Chapters 1 through 8, 10, 11, 12, 14, 17, 18, 19 through 21, 23, 25, through 36, and selected topics from other chapters. Although the arrangement of the text, based heavily on reviewers' suggestions, is from organismal to ecosystem levels, instructors may reverse order of presentation and start the course with Chapters 1 and 2 followed by Chapters 22 through Chapter 25, then Chapters 19 through 21, Chapters 10 through 18, and then Chapters 26 through 36, with selected topics from the other chapters. For classes that need an strong relevancy, instructors may wish to abandon a structured approach for the course or parts of the course by focusing on some current environmental problem and moving from there. For example, in one newscast, CNN discussed the plight of the Florida panther. Its habitat has been badly fragmented. Only 30 to 50 panthers exist and they are separated into small subpopulations occupying patches of highly fragmented habitat. The a total population has lost 75 percent of its diversity, and there is strong evidence of inbreeding: low sperm counts, males with one testicle, and heart-valve defects. Focusing on this situation, the instructor could start off with habitat fragmentation (Chapter 19), follow that with Hardy-Weinberg and population genetics to investigate the effects of inbreeding, minimal population sizes, and problems of captive breeding (Chapter 2). If one assumes that the habitat was restored and animals were introduced into the wild then, the instructor could move to founder effect and bottlenecks, followed by population growth (Chapters 11, 12), food (Chapter 16, predator-prey relationships). Other topics could be handled in a similar manner. The point is the book contains sufficient material to allow extreme flexibility in structuring an ecology course for nonmajors.

# CHAPTER 1

# WHAT IS ECOLOGY?

## Commentary

Chapter 1 defines ecology and sketches its diverse origins, traced in Figure 1.1. If you were to trace the developments of chemistry and physics, for example, you would discover some rather definitive roots which developed into different areas, much like the branching of a tree. But ecology represents the process in reverse. Its beginnings arose from many roots and the sprouts tend to coalesce into a trunk of sorts. For this reason, ecology is difficult to reduce to a limited set of basic principles, and it probably will never develop into a science as exact as physics or chemistry. Biological Nature, so variable and complex, can never be reduced to sets of mathematical formulas as Physical Nature can.

Modern ecology has developed along two major lines, ecosystem ecology and population ecology. Like other sciences, ecology is interested in three basic questions: what, how, and why. Answering such quetions involves experimentation in both laboratory and field. Students should appreciate the scientific method employed in ecological studies and understand why ecology cannot produce as definitive answers and results as more exact sciences. Ecologists can predict probable outcomes of particular actions or events with varying degrees of certainty, which is much better than no prediction at all. Certain outcomes, such as drainage of wetlands will reduce waterfowl populations can be made with a high degree of certainty. We can predict what might happen with global warming, but in spite of simulation models, we cannot say with certainly what the outcome will be. But predictions of potential outcomes should provide a warning not to ignore the threat of such outcomes. Such an understanding of ecology is essential in issues and court cases involving environmental decisions.

The chapter ends with a look at the Application of Ecology. Just as engineering depends upon physics and chemistry, so too does applied ecology depend upon the principles of ecology, especially ecosystem ecology in the management of natural resources and the assessment of and reduction of environmental impacts of human activities. Applied ecology too often has been ignored by certain groups of ecologists who considered it as non-science. That viewpoint is changing rapidly as some theoretical ecologists and many systems ecologists are beginning to see the importance of the application of ecological theory to real world environmental problems, a point that will become more evident as you work through the text.

## Source Materials

The **Selected References** at the end of the chapter provide most of the source material available. Especially valuable are S. Kingsland (1985), R. P. McIntosh (1985), and D. Worster (1977) L. A. Real and J. H. Brown, eds., 1991, FOUNDATIONS OF ECOLOGY: CLASSIC PAPERS WITH COMMENTARIES, Chicago: uUniversity of Chicago Press, provides insights into the conceptual roots of ecology. The development and state of ecology in various parts of the world are well summarised in HANDBOOK OF CONTEMPORARY DEVELOPMENTS IN WORLD ECOLOGY edited by E. J. Kormondy and J. F. McCormick (Greenwood Press, Westport, CN., 1981)

## Discussion Topics

1. The text makes a statement that ecologists cannot give definitive answers. Too what degree is this true? In what situations may ecologists give a relatively definitive prediction? Why may some definitive conclusions for the short term be negated in the long term?

2. To what extent are Environmental Impact Statements change project plans to protect sensitive environments; or are they prepared, them highly modified or ignored because of politicl pressure?

3. Obtain some recent EISs and study them for their content and what say or do not say.

PART II

INTRODUCTION

Natural Selection

## Commentary

The Introduction is a short essay on Charles Darwin, his voyage aboard the The Beagle, the development of his theory of natural selection and evolution, and his interactions with Alfred Wallace. Most students will be familiar with Darwin's name, but most will have a very little appreciation and understanding of the Darwinian theory of evolution. At best their knowledge may be highly distorted. You should urge students to read at least two works of Darwin, THE VOYAGE OF THE BEAGLE and THE ORIGINS OF SPECIES. A number of editions, both paperback and hardcover, are available. Students who cannot be motivated to read through the full length original should be directed to the THE ILLUSTRATED ORIGIN OF SPECIES by Charles Darwin, Abridged and Introduced by Richard E. Leakey (Hill and Wang, New York). To flavor some of the Darwinian theory, students should read some of the volumes by Stephen Gould, such as EVER SINCE DARWIN.

## CHAPTER 2

## NATURAL SELECTION AND EVOLUTION

## Commentary

"Why should we be concerned with natural selection and evolution in an ecology course?" This question may be a reaction of many students. The answers rests in one definition of evolution: the adaptative modifications of organisms over time. Thta is the reason plants and animals are adapted to the environments in which they live; that is why insect pests become resistant to pesticides and weeds to herbicides. Natural selection and evolution are at the heart of ecology and some understanding of those concepts will answer questions that will arise later. So this chapter introduces an evolutionary approach to ecology.

The chapter opens with a brief survey of the geological periods with emphasis on life as it emerged and changed through time.. Considered, too, is the influence of continental drift on the distribution of life. The section concludes with a consideration of the Pleistocene which seem to have been

8

crucial to the spread of humans and the extinction of certain types of mammalian life. Some emphasis should be given to continental drift. .Some students may believe that continental land masses of today are roughly what they have been in distant past. A understanding of continental drift will provide a better appreciation of distribution of life and climate change in geological past.

This brief overview of the evolution of life over geological time leads to a consideration of natural selection. The points to be emphasized here are the differences between adaptation, fitness, natural selection, and evolution. You may need to correct the erroneous idea about natural selection held by some students that the term means the survival of the fittest by the way of fang, claw, and violent struggle.

Some students may wonder how selection can be imposed upon individual sufficiently to change a species or a population. To explain the process you can discuss artificial selection used to improve domestic plants and animals. By selecting and breeding for certain wanted traits and characteristics and by selecting against or eliminating those unwanted plants and animals from the breeding population, humans have been able to develop a wide variety of domestic plants and animals. Consider the various breeds of dogs, breeds of livestock, and improved varieties of cereal grains. Milk production in dairy cattle cattle has been increased greatly by retaining only the best producing cows and breeding them only to bulls with proven ability to transmit high milk production.

Once you have established an understanding of natural selection, you can show how selection operates through genetic variation. The problem here to get the basic ideas across without getting involved in too much genetics. Emphasize the various sources of variation, the role of mutation (including a refutation of of major misconceptions of mutation, thanks to science fiction) and the role of certain types of genetic variation not subject to natural selection. Recent discoveries of molecular genetics is forcing some rethinking about Darwinian selection.

The types of selection are fairly easy for students to comprehend, but they should be cautioned not to arrive at a false conclusion that evolution is a mechanism for constant improvement of a species. Even artificial selection in domestic plants and animals does not result in superior animals. Certain traits may be sacrificed for other traits desired by humans at the time. An examination of the changes in the types of domestic hogs or chickens over the past 100 years points out that fact. The idea that evolution is constantly improving a species should be discarded in students' minds. What natural selection is doing is selecting for the most suitable traits among those available in the genotype. Comprehension of this topic is important to the understanding of the problems of pest control discussed in Chapter 18.

Perhaps the most important section in the chapter from an ecological viewpoint deals with constraints on natural selection. Discussed are founder effect and genetic drift, effects of inbreeding in small populations, viable population size. These topics, once mostly of theoretical interest, have assumed

9

great importance as wild populations of plants and animals are becoming fragmented and isolated through habitat destruction. The result is a loss in gene exchange, in genetic variability. The implications of this loss should be brought to the students' attention relative to the maintenance of species and the restoration of endangered species. This topic, which is at the heart of conservation biology, relates to Chapter 20, which discusses habitat fragment which results in small isolated populations of a species. Interest in role and importance of genetics in the management of wild populations of plants and animals and in the maintenance of genetic diversity in both domestic and wild organisms, in other words, genetic conservation, is increasing rapidly. Two good references are Frankel and Soule, 1981, and Schonewald-Cox et al, 1983, listed under source material, as well as papers in the journal Conservation Biology.

## Source Material

Additional information on geological periods, the evolution of life, and continental drift is available in a number of books on historical geology and biogeography. A good readable discussion of continental drift and its effects on biogeography is C. B. Cox, I. N. Healey, and P. D. Moore, 1975, BIOGEOGRAPHY: AN ECOLOGICAL AND EVOLUTIONARY APPROACH, Blackwell, Oxford, England. An excellent discussion appears in J. H. Brown and A. C. Gibson, 1983, BIOGEOGRAPHY, C. V. Mosby Company, St. Louis, MO. An early exposition of the theory can be read in A. Wegener, THE ORIGINS OF CONTINENTS AND OCEANS (trans. by J. Biram from the 4th rev. ed., 1929), Dover Publications, New York. A good brief article is "Continental drift" by J. T. Wilson in TOPICS IN THE STUDY OF BIOLOGY, Harper & Row, New York, 1971, pp. 382-386. A well illustrated article is "This changing earth" by S. W. Matthews in National Geographic, Vol. 143, No. 1, Jan. 1973 , pp 1-37.

For general background material on genetics see two excellent short texts designed for self study: E. O. Wilson and W. H. Bossert, 1971, A PRIMER OF POPULATION BIOLOGY and D. L. Hartl, 1981, A PRIMER OF POPULATION GENETICS, Sinauer Associates, Sunderland, MA. .An excellent introduction to evolutionary biology which discusses all the ideas contained in this and Chapter 3 is D. J. Futuyma, 1984, EVOLUTIONARY BIOLOGY, Sinauer, Sunderland, MA. The problems of generic drift role of founder principle, and the lack of genetic diversity in natural populations is discussed in three important books, C. M. Schonewald-Cox et. al. (eds.), 1983, GENETICS AND CONSERVATION, Benjamin/Cummings, Melno Park Ca.; and O. H. Frankel and M. E. Soule, 1981, CONSERVATION AND EVOLUTION, Cambridge University Press, Cambridge, England, M. Soule (ed.) 1986, VIABLE POPULATIONS FOR CONSERVATION. Papers on current problems and issues and research in conservation biology appear in the journal Conservation Biology.

## Discussion Questions

1. Considering all the assumptions it has to meet, of what value is the Hardy-Wienberg Law?

2. Consider the statement "Regardless of the kind of selection, the characteristics selected are not necessarily the best of all possible traits, but

10

rather the most suitable of those available." Isn't evolution supposed to result in the continual improvement of a species?

3. Why can't natural selection prepare a population for the future?

4. Does selection operate on the phenotype or the genotype?

5. Do the "best" genes necessarily prevail over all others? If so, how can you account for genetic variation?

6. What are "inferior" genotypes? Does it mean inferior for the particular environment in which the populations exists? What if the environment changes and the "inferior" genotypes survive and the "superior" genotypes perish?

7. What are the problems faced by a species whose populations are reduced to isolated fragments? Consider here, too, the problem of fragmentation of habitat discussed in Chapter 20.

8. What genetic problems are involved in attempts to rescue a species from extinction. Consider the whooping crane, the black rhino, black-footed ferret, European bison, the lion tamarin. For information see Schonewald-Cox et. el., 1983; Frankel and Soule, 1983; and M. E. Soule and B. A. Wilcox, (eds.), 1980, CONSERVATION BIOLOGY, Sinauer, Sunderland, MA, and M. Soule (ed), 1986, Examples of two case history papers are C. Packer et al., 1991, Case study of a population bottleneck: Lions of Ngorongoro Crater, Conservation Biology 5:219-230; and M. V. Ashley et al., 1990, Conservation genetics of the black rhinocerous (Diceros bicornus), I: Evidence from the mitochondrial DNA of three populations, Conservation Biology 4:71-77.

# CHAPTER 3

## SPECIES AND SPECIATION

### Commentary

The subject of natural selection leads easily into speciation. Students may question why they should be concerned with this topic. The is because it is one of environmental concern. The species concept is at the heart of the endangered species controversies and recovery programs. For example why should we be concerned about the extinction of the dusky seaside sparrow when other subspecies exist? The species is not extinct.

With an emphasis on molecular biology in general biology courses, many students have little knowledge of taxonomy and the classification of organisms. For this reason I have emphasized the classification scheme in Box 3.1. All students should be able to classify a given organism from Kingdom through Species. Point out the fact that classification of organisms differs from naming organisms. The naming of organisms which has a definite set of rules for both plants and animals is nomenclature.

The idea of the morphological species is familiar to most students. They are using the concept each time they seek the identity of an organism in a field guide. They are in a way familiar with the idea of a biological species until they are encouraged to look at the concept critically and have to arrive at some decision about asexual organisms. A discussion of the problems encountered trying to associate all living organisms within the framework of a biological species should emphasize the problems biologists face in trying to force great variety of life into human classification schemes.

A distribution map of some selected species of animals showing the array of "well-defined" subspecies, for example ruffed grouse, cottontail rabbit, or white-tailed deer, could be the basis of a discussion of geographical variation in species. The distribution will be more informative if morphological data on the subspecies are available. From such information students might be able to arrive at some conclusion about the presence or absence of a cline in the species. Contrast a cline with an ecotype and with a geographic isolate.

The process of speciation comprises most of this chapter. Allopatric or geographic speciation is the classical type, and, some will argue, the only type of speciation. But it is difficult to comprehend how all the enormous number of species on Earth, especially among the insects, could all have arisen only by sympatric speciation. Another form is parapatric speciation which in a way can be considered as speciation on microgeographic scale, because the incipient species are isolated by some environmental situation. The most controversial type of speciation is sympatric. Is polyploidy, for example, an instance of sympatric speciation? Sympatric speciation might result from genetic changes

12

outside of natural selection. Or can sympatric speciation result from even more restricted type of geographic isolation such as restriction to certain host plants or animals? The question of speciation should stimulate some interesting discussions.

There are numerous examples of adaptive radiation in plant and animal kingdoms. You could have students seek out more examples of convergence. The idea of character displacement is considerably more nebulous and a questionable concept.

Some problems of speciation are discussed in a series of papers in ANNUAL REVIEW OF ECOLOGY AND SYSTEMATICS, 15, 1984, and in the book edited by D. Otte and J. Endler (eds.), 1989 SPECIATION AND ITS CONSEQUENCES, Sinauer, Sunderland, MA. Both are excellent sources of additional material on the subject.

Some newer ideas relative to evolution and speciation are gradualism versus punctuated equilibrium and the idea of instantaneous speciation. But emphasize that the term instantaneous is relative and involved thousands rather than millions of years. A nicely illustrated idea of the concept appears in Stebbins (1982:18),using as an example the evolution of elephants. Another concept considers natural selection at the species level. Certain species are more vulnerable to extinction than others and their loss results in the removal of a species rather than genes within a species. Some discussions could center about the characteristics of extinction-prone species and characteristics of species that so far seem be successful in escaping extinction. What for example are the characteristics that resulted in the extinction of the ivory-billed woodpecker and allowed the large pileated woodpecker to occupy successfully a range of habitats from deep wood to wooded suburbia and small cities? Or in the decline of the New England cottontail rabbit and the success of the eastern cottontail rabbit?

The common conception of evolution is that it takes place over centuries This belief arises because students fail to remember that evolution involves the change of gene frequencies over time. Evolutionary changes can take place rapidly. The English sparrow, for example, evolved into clinal races across North America in less than 75 years. The introduced ring-necked pheasant is a much different bird in North America than its original Mongolian ancestors.

## Source Materials

The list of suggested readings at the end of Chapter 3 in the text presents major source materials. Students should also be directed to Stephen J. Gould, 1989, WONDERFUL LIFE: THE BURGESS SHALE AND THE NATURE OF HISTORY for a superb discussion of evolution and speciation.

## Discussion Topics

1. Classification and systematics are based on the judgement of authorities who have studied various taxons in detail. In 1982 the American Ornithologists Union revised the classification scheme of North American birds, particularly at the

family and subfamily levels, the details of which appear in the 7th edition of THE CLASSIFICATION OF NORTH AMERICAN BIRDS. Compare the changes by consulting old bird guides (most still follow the old classification) and the latest checklist. Refer to the literature cited to discover why such changes were made.

2. Suggest why such species as the horseshoe crab and the coelacanth have evolved so slowly, changing little from their fossil relatives, compared to other species in sea.

3. One means by which sympatric speciation might occur is if individuals in population are homozygous for a trait that enables them to exploit a different host, food, or microhabitat. For example, if A utilizes small red fruit on plant species 1 while A' genetically is able to utilize the large red fruit of plant species 2 exclusively, could you not consider that geographic isolation on a microscale, even though the two, A and A' live in the same macrohabitat and are therefore sympatric? What constitutes geographic isolation?

4. There is some question about the amount of gene flow among local populations. Over what distance does genetic interchange take place? Consider two populations of white-footed mice which are separated by several miles and two valleys. How much interchange do you expect could take place? If interchange is minimal to non-existent, then why don't the two populations evolve into separate species? (This question relates back to Chapter 2.)

5. Have students consider question 16. Problems associated with many endangered species are discussed in excellent case history studies that appear in AUDUBON WILDLIFE REPORTS, Academic Press, Orlando, FL. These case histories provide a range of species to consider.

# PART III

## INTRODUCTION

### Homeostasis and Systems

#### Commentary

Part III is ecology at the organismal level, concerned with the responses of organisms to their physical environment. The introduction to Part III introduces several concepts important to the material covered. One of these is the Law of the Minimum. It is a law often misunderstood or taken too much at face value. As stated, it applies only to equilibrium conditions. All other aspects held constant, then the response of an organism is controlled by the environmental material in the shortest supply. If that limiting material is increased, then organism may so function that some some other material may become limiting. Thus the law really relates to the concentrations of all resources present.

In some situations the amount of a resource may be far in excess of the organisms needs, or it may interact synergistically with others present. In such situations the material may become toxic to the organism. Then another law formulated by Black may apply, the Law of the Maximum. Victor Shelford considered both laws and came up with the Law of Tolerance. Organisms are limited in their distribution, growth, and reproduction by their degree of tolerance to any one of a number of conditions or to conditions combined. Tolerance may vary with stage of life history. Some organisms, for example, could survive as adults under conditions in which they could not reproduce successfully or in which the young could not survive. Tolerance may vary among the life stages of an individual, or seasonally within individuals, which brings up the idea of acclimatization. The prefixes eury- and steno- indicate wide and narrow ranges of tolerance respectively. Thus an organism may have a narrow tolerance for one condition and a wide tolerance for others. It distribution or fitness would be limited by the condition for which it had the least tolerance. Such points should be emphasized, because they are fundamental to a student's understanding why plants and animals respond as they do to pollution and other environmental changes.

Another concept is homeostasis, a physiological term referring to the ability of organisms through a feedback mechanism to maintain a physiological balance. Emphasize set point, positive feedback, negative feedback, and homeostatic plateau, all of which make up a cybernetic system. The idea of cybernetic systems has been applied to population and ecosystem levels as compartment models which act in a cybernetic way. These systems include inputs and outputs, compartments or boxes, and driving forces among them. This systems approach is used through the text.

15

# CHAPTER 4

## CLIMATE

### Commentary

Chapter 4 is basic to much of the text. The Sun is Earth's major source of energy. Sunlight energizes photosynthesis. Its thermal energy heats land and water, drives the local and global movements of air and ocean currents, further influenced by the spinning of Earth. This spinning produces the Coriolis effect.

Students should gain some knowledge of atmospheric movements brought about by the heating and cooling of air masses. Then they will better appreciate such observable phenomena as morning and valley fog, inversions, air stagnation, night breezes, upslope and downslope winds, all ecologically important.

While broad weather patterns may influence human activity, local or microclimates are much more important to populations of plants and animals. Micro- or little climates vary across the local landscape according to slope position, soil, vegetation, and microrelief such as depressions or rises in the ground. These microclimates, in part, produce the patchy distribution of plants and animals.

Study of microclimates in detail might be postponed until after the study of Chapters 5, 6, and 7, because microclimatic differences relate to changes in temperature, moisture, and light. Perhaps the discussion of microclimates should have followed those chapters, but that would have created some other organizational problems in this text. That is the problem with trying to organize ecology. All things are interrelated and one has difficulty deciding what to discuss first.

Although the emphasis on microclimates is more on the natural environment, urban areas have their own microclimates. Pavements and building hold heat and reflect heat after sundown. Temperature differences within the city influence wind movements. Shadows cast by buildings create north and south exposures. So the city environment does provide opportunities to observe microclimatic effects on the growth and distribution of plants, cultivated and otherwise, and on the movements and distributions of animal life, particularly urban birds.

Human activity has a pronounced effect on local, regional, and global weather patterns, and in turn affecting vegetational pattern and distribution. Of particular concern is global warming and the greenhouse effect, introduced here and considered again later in the text. Some discussion of both should be introduced at this point.

## Source Materials

**Selected References** at the end of the chapter provide excellent sources. Although showing its age, R. Geiger, 1965, CLIMATE NEAR THE GROUND, Cambridge, MA: Harvard University Press, is still the classic reference on microclimate. Less detailed is R. Lee, 1978, FOREST MICROCLIMATOLOGY, New York: Columbia University Press. A good introduction to weather is D. Ludlum, 1990, AUDUBON SOCIETY FIELD GUIDE TO NORTH AMERICA WEATHER, New York: Knopf. The most lucid discussion of weather patterns on a regional and local basis is a handbook on the subject written largely for Forest Service personnel. It is FIRE WEATHER, U. S. D. A. Agricultural Handbook No. 260, by M. J. Schroeder and C. C. Buck. I believe it is now out-of-print, but you may query the Government Printing Office. It is available in many libraries. Two excellent introductions to global warming are S. H. Schneider, 1989, GLOBAL WARMING, San Francisco: Sierra Club Books; and M. Oppenheimer and R. H. Boyle, 1990, DEAD HEAT: THE RACE AGAINST THE GREENHOUSE EFFECT, New York: Basic Books.

## Discussion Topics

Because the material in this chapter lends itself more to discussions of local weather observations and to activities than to questions per se, some of the questions are really field activities.

1. Assume you are camping in mountainous or very hilly terrain. Where would you place your campfire so that smoke will drift into your tent site?

2. Why is there no or little temperature inversion on a cloudy or windy night?

3. Each morning observe indications of air inversions such as morning fog, layer of smog, and so on. What were the night conditions that induced the inversions? Discuss.

4. Note temperature differences beneath a forest canopy and in the open; on the shaded and sunlit side of a building.

5. Observe the movements of animals through the day? What special microclimate situations are they exploiting and why? You don't have to observe wild animals. Dogs, cats, and cattle will do.

6. Question 13 in the **Review and Study Questions** should be given strong consideration as a discussion or report topic.

# CHAPTER 5

## TEMPERATURE

### Commentary

After a brief introduction to the mechanisms in heat transfer between an organism (or any object) and its environment and their relationship to the maintenance of heat balance, the chapter moves on to its major topic, the adaptation of organisms to temperatures of the environment. Much of the chapter concerns endothermy and ectothermy and their relationship to Homeothermy, poikilothermy, and heterothermy.

Most of this chapter is straight forward. A major objective should be an understanding of the terms and the concepts they imply. Emphasize that homeothermy and endothermy and poikilothermy and ectothermy are not synonymous terms. They imply something different. Some homoiotherms at times can be ectothermic and some poikilotherms can be endothermic. Organisms which exhibit some characteristics of each are called heterotherms.

Homeothermy and endothermy relate to body size, which is critical to the maintenance of heat balance. Basal metabolism in homeotherms is proportional to body surface area, about 0.07 ly/min. This is roughly equivalent to the long radiational balance for a surface at air temperature. Some interesting discussions could center about the significance of size in birds, mammals, fish, and insects and other invertebrates.

Knowledge of acclimatization is important in understanding the ecology of poikilotherms, especially aquatic ones. They exhibit a wide range of seasonal tolerances at any one time. What temperatures may be tolerable or optimal at one time of year may be lethal at another, and sudden changes in temperature at any any season can be lethal. This topic is particularly relevant to thermal pollution of rivers and lakes by power plants that draw water from them for cooling and discharge heated water back to the aquatic environment.

There are four areas which students find interesting, in part, I suspect because they can represent the "gee whiz, isn't nature grand" category. Actually the areas represent good examples of evolutionary adaptation to environment stress. One is countercurrent circulation. The second is the use of "antifreeze" solutions by plants and some animals such as fish and amphibians to increase their tolerance to subfreezing temperatures. Here you may wish to emphasize other physiological mechanisms used by plants to increase cold hardiness. The third is temporary endothermism in some plants, especially the skunk cabbage. Students might speculate on the evolution of such endothermism. Was the major selection force the attraction of early flying insects or what? Remember the family to which skunk cabbage belongs is a tropical one. The fourth is hibernation. Hibernation is not well understood, especially by lay public. There is more to hibernation than going into a long, uninterrupted sleep, which is the common conception, or misconception. There are various degrees of the hibernating state, including physiological changes and the depth of the winter sleep. What are the differences between the hibernating states of a black bear and a chipmunk, or between a chipmunk and a bat? Providing a good discussion of hibernation is A. R. French, 1988, The patterns of mammalian hibernation, American Scientist **76**:568-577.

## Source Materials

**Selected References** at the end of the chapter provide access to the general literature. Of those listed the most accessible reference for plant ecophysiology is W. Larcher, 1980, PHYSIOLOGICAL PLANT ECOLOGY 2nd. ed., New York: Springer-Verlag. Temperature regulation by poikilotherms, specifically insects, including heterothermy and endothermy, is nicely discussed by various authors in B. Heinrich, ed., 1981, INSECT THERMOREGULATION, New York: Wiley. Two excellent references for vertebrate animals are G. E. Folk, 1974, TEXTBOOK OF ENVIRONMENTAL PHYSIOLOGY, Philadelphia: Lee & Febiger, which has a strong ecological approach, and K. Schmidt-Nielsen, 1990, ANIMAL PHYSIOLOGY: ADAPTATION AND ENVIRONMENT 4th ed., New York: Cambridge University Press. The role of body size is discussed in a book by the same author and publisher: SCALING: WHY IS ANIMAL SIZE SO IMPORTANT?

## Discussion Topics

1. Questions 17 and 18 of the **Review and Study Questions** provide a basis for discussions and reports. Question 17 could be followed by a study or report on the effects of local power plants on aquatic environments, if water-cooled plants are involved. Seeking the information from both local utilities and state conservation departments can be revealing politically as well as ecologically.

2. Locate a marshy area supporting skunk cabbage. In very early spring take temperature of the soil surrounding and away from the plant, as well as temperatures inside the spadix compared to the surrounding air temperature. Observe for a period of time the insect activity about the plant.

# CHAPTER 6

## MOISTURE

### Commentary

The influences of moisture relate to the physical characteristics of water. Students may be familiar with the physical properties of water from physics and chemistry courses, although they may not be aware of the ecological significance of those properties. Other students may have no concept of the physical structure and behavior of water. These physical characteristics must be emphasized, for they relate to most ecological processes and the structure of aquatic ecosystems.

Animals can respond to moisture deficits and excesses behaviorally, as well as physiologically, and both responses are emphasized. Being fixed in place, plants' responses to drought and flooding are physiological only. Although students recognize the response of plants to drought, they are much less familiar with the response of plants to flooding. An knowledge of observable physiological effects of flooding on plants will increase students' awareness of the impacts of certain human activities such as road construction, dam construction and the like on trees and other plants. They will be better able to interpret the landscape.

Not all moisture problems result from lack of water. Some come about because of saline water, which creates physiological drought. The material here will relate to the discussion of salt marshes in Chapter 36.

Moisture, like temperature, influences the distribution of plants. If you know the moisture requirements of various plant species, you can assess the nature of the physical environment. You would not expect to find cardinal flowers blooming on a dry hillside. The presence of small patches of cattails along interstate highways points out poorly drained areas. But the ultimate distribution of plants and associated animal life relates to the interaction of temperature and moisture. In fact, the two are almost inseparable. In hilly country, for example, mesic situations are usually cooler places where evapotranspiration is reduced compared to warmer xeric situations. The interrelationships are most pronounced in regional and global distribution of vegetation. Students' familiarity with this interaction will increase their awareness of vegetational patterns and animal distribution across the landscape.

### Source Materials

The physiology texts cited in the Source Materials for Chapter 5 also cover material on moisture and water economy. Additional references listed at the end of the book chapter provide more detailed information on plant responses to moisture. A well-written general introduction to water is C. A. Hunt and R. M.

Garrels, 1972, WATER, THE WEB OF LIFE, New York: Norton.

## Discussion Topics

1. Preparation of climographs for the local area, regions, different parts of the state or the country, will emphasize for the students the climatic differences between regions and lead to a better understanding of the differences between say the humid eastern North America and the arid southwest. The differences relate directly to problems of water shortages, settlement and the like. Climographs make a good exercise in environmental geography. Data are available in federal weather bureau monthly summaries available at most libraries.

2. What are the microclimatic reasons why forest tree growth in temperate regions is greater on north-facing than on south-facing slopes?

# CHAPTER 7

## LIGHT

### Commentary

Following a brief review of the nature of light, the discussion of light is divided into two parts. The first discusses light as an environmental influence on plants, its diminution in the plant canopy and in water and plants' responses to that diminution. Some plants can grow and reproduce in low light intensities; others cannot survive at all. For some plants, too much light is harmful; for others too little is detrimental. This response is expressed as shade tolerance.

Many students may be aware of shade tolerance and intolerance as it relates to house plants and garden flowers. To open the topic to discussion, you might call upon students to list some responses of plants to light and shade based on their own observations. The concept of shade tolerance and intolerance is an important background for topics to come, particularly opportunistic and equilibrium species, community structure, and succession.

The second part of the chapter deals with photosynthesis. This topic could be incorporated with Chapter 21, Energy Flow, but in this revision I choose to introduce photosynthesis here, because it also relates to shade tolerance and intolerance, as well as a response to light. Many students will be familiar with the basic photosynthetic process. They may know little about the $C_3$, $C_4$, and CAM cycles. The cycles are discussed here in sufficient detail for the students to appreciate the differences between them. The major emphasis is on the ecological significance and importance of them.

### Source Materials

Shade tolerance in plants is well discussed in both Larcher and Etherington (see Source Materials, Chapter 5), but most extensive treatment is in J. Grime, 1971, PLANT STRATEGIES AND VEGETATIVE PROCESSES, New York: Wiley. Another good but somewhat dated reference is R. Bainbridge, G. C. Evans, and O. Rackham, eds., 1966, LIGHT AS AN ECOLOGICAL FACTOR, Oxford, England: Blackwell.

### Discussion Topics

1. What is the relative ranking in shade tolerance for forest trees and shrubs? Prepare a tolerance table for local forest trees and shrubs. Refer to S. H. Spurr and B. V. Barnes, 1980, FOREST ECOLOGY, New York: Wiley; and R. M. Burns and B. H. Honkala, 1990, SILVICS OF NORTH AMERICA, Vol. 1, CONIFERS; Vol. 2, HARDWOODS, Agricultural Handbook 654, Washington, DC: U. S. Department of Agriculture, Forest Service. The latter work is an indispensable addition to the ecological library.

# CHAPTER 8

## PERIODICITY

### Commentary

Circadian rhythms and photoperiodism receive only passing comment in most ecology texts, yet the role of light and biological clocks in all living organisms is one of the most important aspects of ecology. They influence such daily activities as song, foraging, periods of activity and inactivity, opening and closing of flowers, and such seasonal activities as breeding seasons, migration, hibernation, periods of blooming in flowers. These time keepers maintain the living world in synchrony with the night and day and the seasons.

The importance of light in controlling seasonal activities through changing daylength is demonstrated not only in the behavior of plants and animals in the wild but also in the manipulation of breeding and flowering in domesticated animals and plants. Lights are kept on in poultry house to maintain a standard daylength to insure the continuation of egg laying by hens. Florist bring chrysanthemums, roses, and other plants into bloom to maintain a steady supply of cut flowers of all types through the year.

Humans, too, are under the control of a biological clock. Students' own experiences can serve as examples. Hours of waking and sleeping are set by student's own daily regime. Day people run on a different clock than night people, who initially have a difficult time of adjusting to a night day activity period when a day night regime is normal. Jet lag is another good example of an interference with a set biological clock. Adjusting to a new time zone requires a phase shift in humans, just as it does in other animals. That involves some physiological discomfort until the phase shift is completed. Circadian rhythms in physiological mechanisms is important to the medical profession, as doctors attempt to handle surgery and other treatments to certain daily phases of physiological circadian rhythms.

Certain other time keepers are important in specialized environments, particularly the marine environments where activities of organisms are influenced by tidal rhythms. In tropical environments, where day and night are of equal length, and in desert environments other seasonal time cues are important, particularly rainfall or wet and dry seasons.

### Source Materials

Literature on biological clocks is extensive. Recommended introductions into the literature are given in the the list of **Selected References**. Students will find Saunders, 1982 and Winfree, 1986 good sources of information. For an intriguing introduction to biological clock in humans consult a book missing from Selected References: M. C. Moore-Ede, F. S. Sulzman, and C. A. Fuller, 1982, CLOCKS

THAT TIME US: PHYSIOLOGY OF THE CIRCADIAN TIMING SYSTEM,
Cambridge, MA: Harvard University Press.

## Discussion Topics

1. Prepare a graph of the waking times of selected local birds, such as robins
and cardinals, marked by first morning songs, against time of local sunrise. Plot
the waking times against curve for sunrise. What relationship exists? (For the
dedicated only; this project requires early rising, especially as the spring wears
on.). What conclusions can you draw from the graph. How do your daily
observations relate to seasonal periodicity?

2. Develop a chart to show how much an individual's biological clock is out of
synchrony with the local environment when one flies to Paris, Moscow, Hong
Kong, Sidney, Australia and when on returns.

# CHAPTER 9

## SOIL

### Commentary

Few students have any real concept of soil. To many of them soil is dirt, the stuff shoved up by bulldozers at construction sites, something in which you plant some flowers. Too few appreciate the fact that soil is the basis of terrestrial life, the foundation upon which we exist. For such a vital component of Earth's ecosystem we take little care of it, depleting its nutrients, poisoning it with chemicals, allowing it to be eroded away by water and wind, burying it beneath concrete and asphalt, removing millions of acres of it from any sort of production. Because few students take any course on soils, a few hours should be spent on the topic. Because a solid background in soil science is usually restricted to those who study agricultural sciences, forestry, or resource management, many who teach general ecology may have limited knowledge of the subject. If this is the case, don't gloss over the subject. Call on help. An excellent source is the Soil Conservation Service local, regional, or state offices. You may be able to arrange a lecture and field trip on soils with them. Or if you are located at or near a land grant university, call upon the Agronomy or Soils Departments.

If field trips are possible, students will benefit from some actual experience of soils in place. Such trips, of course, will require the assistance of someone knowledgeable in regional and local soils. Soil profiles exposed along roadside cuts, or dug in woods and grassland will provide visual experience with soil horizons. Have students determine the soil texture and pH of each horizon. Students can make some inferences on the relationship of pH to calcium and calcium recycling. You can also have the students investigate differences in in soil horizons beneath a coniferous and deciduous forest stand. Studies can be extended to include the extraction of soil organisms from litter layers of different ecosystems: grassland, coniferous forest and deciduous forest. You may also have the students explore the relationship between soils and vegetation patterns. Essential for such a project would be a local soils map. The soils of most of the United States have been mapped (and remapped). Consult you local or state Soil Conservation Service office for the availability of maps and manual for your area. Local soil surveys are an indispensable part of a working ecological and environmental library.

The chapter ends with a discussion of soil erosion. This topic should be given some emphasis. Soil erosion is evident after every heavy rain and spring snow melt, as muddy waters gush along roadsides, roll off construction sites and crop fields, flow from storm drains to muddy streams, rivers, and lakes. Students should be aware of the magnitude and consequences of such erosion, especially in their own local areas.

## Source Materials

Among the **Selected References** two books should be part of an ecological and environmental science library. One is a basic reference that should be part of an ecological library is H. Jenny, 1980, THE SOIL RESOURCE; the other is N. C. Brady, 1990, THE NATURE AND PROPERTIES OF SOIL . For life in the soil see SOIL BIOLOGY published by UNESCO (Paris: UNESCO) and J. A. Wallwork, 1973, ECOLOGY OF SOIL ANIMALS, New York: McGraw-Hill. The latter, unfortunately, is out of print. Experiencing the same fate is Peter Farb, 1959, THE LIVING EARTH. It is a beautifully written introduction to soils. For a visually rewarding look at soil, its nature, inhabitants and problems see National Geographic, Vol. 166, No. 3, September 1984, pp. 350-388.

## Discussion Topics

1. A common misconception is that soil can be kept in a high state of productivity by adding commercial fertilizer to replace nutrients withdrawn from the soil. What is the fallacy of this attitude? What about organic matter and activity of soil organisms? Why is there an increasing interest in organic farming.

2. Particular attention should be given to **Review and Study Questions** 14, 15, and 16.

# Part IV

# POPULATION ECOLOGY

## PART IV

### INTRODUCTION

#### Populations

#### Commentary

The short introduction covers material that does not fit well into the chapter proper. In it you will find a definition of population, attributes of populations, and special considerations that differentiate plant populations from animal populations.

Because of the growing importance of demographic principles to the study of plant populations, it is important that students understand the differences. Plant populations do not fit all of the parameters applicable to animal populations. Is a plant an individual or is it an assemblage of populations of buds, leaves, flowers, twigs, each with its own natality and mortality rates? What about a group of sprouts arising from a cut stump or a number of root suckers. Is a clump or sumac or aspen, or a patch of strawberries a group of individuals or are they parts of an individual? Be sure students are familiar with the terms ramet and genet. They will come up again.

#### Source Materials

The best reference on the nature of plant populations is J. L. Harper, 1977, POPULATION BIOLOGY OF PLANTS, London: Academic Press. A later reference addressing the questions posed above is R. Dirzo and J. Sarukhan, 1984, PERSPECTIVES ON PLANT POPULATION ECOLOGY, Sunderland, MA: Sinauer Associates.

#### Discussion Topics

1. Consider the questions presented in the commentary above.

2. Discuss the idea that metamerism in plant, the production of genets by ramets, is a means of spreading the risk of a ramet's extinction.

# CHAPTER 10

## DENSITY, DISTRIBUTION, AND AGE

Some ecologists question why density should be considered in an ecology text, questioning its evolutionary significance. I have tried to answer that question in the section on density. It is an elusive property, but nevertheless an important one, especially in the management of exploited plant and animal populations. Contrast crude and ecological density.

How are organisms distributed or dispersed in space and time? To stimulate an answer to that question, have students consider classroom use in the building during the day, with its changing size and distribution of students both in space and time. The idea of patchy environment is introduced here. The concept will reappear a number of times later in the text. Methods of determining density are discussed briefly. Interested students should be directed to books recommended in **Source Materials.**

Age structure is a useful population parameter in population studies. For example, from life table information one can construct a pyramid depicting a stationary age distribution - the ratio of age classes in a population which has reached a constant size and in which birth rate equals the death rate. That, of course, is a hypothetical situation, but nevertheless does present a yardstick against which changes in the age structure of a population can be observed. If used carefully, age structure is of some value in wildlife management.

Age structure can tell you something about a population if you use it with care. A series of age pyramids enables you to follow the fates of various age classes over time. These changes can be associated with environmental changes in the case of wild species and economic and historical situations (such as war) in human populations. Because of unforeseen events, age structure may be a questionable means of predicting future population trends.

Basic types of age pyramids are given in Figure 10.6. Examples of typical age pyramids follow. Figure 10.10 provide examples of "age pyramids" for two trees, balsam fir and sessile oak. Note that the age pyramid for balsam fir involve diameters rather than age. The reason is that most stands of plants, trees in particular, consist of populations of even-aged individuals. Size differences, too often regarded as reflecting age (smaller individuals being the younger), reflect instead intra- and interspecific competition among individuals. The "young" trees are actually suppressed individuals of the same age. Note that large individuals, the dominants, are much fewer in number than small individuals, giving the impression of large base of young individuals. The age pyramid for oak provides no indication of size.

## Source Materials

For information relating to age structure and its use in human populations see N. Keyfriz and W. Flieger, 1971, POPULATIONS: FACTS AND METHODS OF DEMOGRAPHY, San Francisco: W. Freeman. For estimating population densities and other parameters see the books listed under **Selected References** in the text.

## Discussion Topics

1. Investigate the population history of your town, city, or region. assign different census years to various students or student groups, have them consult regional census data for various years, and construct age pyramids. Then discuss what the data reveal. Follow the age classes over the decades.

2. Have students determine the pattern of human population distribution in your area. What is the degree of clumping? What landscape features influence population distribution?

3. How is the population of your area divided into voting districts? How are the boundaries determined. What is the problem of equalization?

Note: although these questions relate to human populations, the projects relate directly to study of natural populations. The problems which appear in the above studies are magnified many times when you investigate natural populations.

# CHAPTER 11

## MORTALITY, NATALITY, AND SURVIVORSHIP

### Commentary

At this point, it should be evident that a close relationship exists between age structure and mortality and natality, the subject of this chapter. Mortality comes first because death influences the number of surviving organisms that can reproduce. Stress the positive side of mortality - survivorship. An account of survivorship is tabulated in a life table (often called instead an account book of death). How much time you want to spend on life tables is your decision. You may or may not want students to construct a life table. for which I give no directions, but see Source Materials.

Students should be aware of certain facts about life tables. They apply only to the population at the time the data were collected. As population parameters change, so will the life table. Life tables are static, reflecting a population with a stationary age structure and zero growth in which births equal deaths. This does not imply that life tables are of little value. They are useful for comparing a given population through time or to compare one population with another. They also provide insights in various aspects of mortality, fecundity, reproductive values, life expectancy, and the like. Both mortality curves and survivorship curves are plotted from data in the life table. Mortality curves plot rate of mortality, the $q_x$ column, against age; survivorship curves plot survivors, the $l_x$ column, against age.

This chapter can easily give one the impression that I overemphasized mortality and survivorship curves. But I have my reasons. If you look at population biology and ecology texts, you will discover that beyond the basic theoretical curves, as illustrated in Figure 11.2, they provide little in the way of real life survivorship and mortality curves. I have provided examples of the three types of curves for various organisms from plants to invertebrates to vertebrates. Note that for plants survivorship curves can be constructed not only for populations of individual plants, but for metapopulations of leaves and other parts as well.

If you can obtain age-specific fecundity for the females of a population, then you can construct fecundity tables for a population which relate births with survivorship among the various age classes. And from the $l_x m_x$ column, the average number of births per individual, one can plot a fecundity curve (see Figure 14.1) for the population, which visualizes the pattern of births among females of the population. From the fecundity table one can also calculate the reproductive value of the different age classes, important information for the management of exploited and pest populations. From it one can also determine the net reproductive rate, $R_o$, useful in constructing growth curves for a population, discussed in Chapter 12. Note that the net reproductive rate

31

considers only the female portion of the population.

## Source Materials

Source materials for this chapter include those for Chapter 10, as well as those listed in the text.

## Discussion Topics

1. Compare mortality, survivorship, natality , and fecundity of a plant population with an animal population. Where do the differences lie?

2. I did not provide a life table for humans. Students may seek the latest ( which is usually not very recent) from several U. S. Government sources, such as Vital Statistics. A generalized life table is not very informative because life table data and thus life expectancy vary among the races. White males and females, for example, have a higher life expectancy than black males and females; and of these groups black males have the lowest life expectancy. As a further exercise, students may seek out life tables for other countries.

# CHAPTER 12

## POPULATION GROWTH

Chapter 11 ends with the calculation of the net reproductive rate, $R_o$. Chapter 12 begins with the same, using the net reproductive rate to estimate the rate at which populations grow. The use of $R_o$ to calculate population growth has some pedagogical advantages over r, rate of increase. But once students have seen how $R_o$ fits into the exponential growth equation, they should be introduced to r. Rate of increase as presented in this chapter is only an estimation; a more precise estimation is through the use of the Euler equation, as demonstrated in Student Resource Manual accompanying R. L. Smith, ECOLOGY AND FIELD BIOLOGY, 4th ed. and in C. Krebs, 1985, ECOLOGY, p. 190.

Exponential growth represents a population growing at rate of compound interest. Point out that the exponential growth equation is the same as that for calculating how fast an initial sum of money will grow at some specified interest rate over time. Populations growing exponentially exhibit stable age distribution in which the age distribution remains proportionately constant.

Populations don't exhibit exponential growth for long. As population density increases, population growth slows down. Theoretically population growth will decline to 0, births will equal deaths, and the population will arrive at an an asympote, K, the carrying capacity of the environment. This growth may produce a sigmoidal growth curve, but rarely. Only a laboratory population under highly controlled conditions even approach true sigmoidal growth. The value of the logistic equation is its use as a model to test natural population growth.

Growth of wild populations introduced into new unfilled habitat at first is more or less exponential; the population reaches a peak, then declines, often sharply, as resources diminish. As the population spreads into new areas, the same pattern follows. The population grows rapidly in response to abundant resources. In areas colonized early, the original center of invasion, the population becomes adjusted to lower level of resources, he result of changes brought about by the original eruptive outbreak or exponential growth of the population. Over time the population sizes will fluctuate, increasing and decreasing as resources and environmental conditions fluctuate, with a time lag.

Slowing of population growth often is attributed to the so-called environmental resistance, an old-fashioned term (introduced by the animal ecologist R. N. Chapman in 1928) that refuses to die and continues to surface in some biology and ecology texts. Population growth slows because of both intrinsic and extrinsic pressures on the population including behavioral interactions, resource scarcity, nutritional deficiencies and the like. The environment is not actively resisting population growth.

The idea of carrying capacity, too, has been carried to the extreme and worked to death in some applied areas of ecology. Too often students and later in life as biologists, have the idea that $\underline{K}$ is some magical number, some upper limit beyond which populations cannot expand in a given habitat. Emphasize that $\underline{K}$ is an abstraction, a theoretical equilibrium density toward which populations tend to grow and about which they tend to fluctuate. This tendency comes about by density-dependent mortality or reduction in reproduction. At this point distinguish between a population fluctuation and a cycle. Too many call any fluctuation between population highs and lows as cycles.

The last section concerns extinction, a situation in which deaths exceed births and $\underline{r}$ is negative. Extinction is an insidious process. Rarely does it occur simultaneously over a species range. It begins as local extinctions which proceed to a point where the species populations become fragmented and isolated with all the attendant problems of restricted gene flow and genetic drift and loss of fitness. Refer back to Chapter 2. Some species may be more extinction-prone than others, but the only cause of the extinction of species in historical times is habitat alteration or destruction by human hands. The process of extinction by human agencies seems to be increasing exponentially.

## Source Materials

References given in sources for Chapter 10 , 11, in addition to those presented in Selected References provide more than adequate material on population growth. Extinction is discussed in detail in O. H. Frankel and M. E. Soule, 1981, CONSERVATION AND EVOLUTION, Cambridge, England: Cambridge University Pres and in M. H. Nitecki, ed., 1984, EXTINCTIONS, Chicago: University of Chicago Press.

## Discussion Topics

1. Carrying capacity, $\underline{K}$, in most population work refers to the average upper limit a population experiences over time. But carrying capacity has other shades of meaning and is used in different contexts in population management onsult R . F. Dasmann, 1964 (or later editions), WILDLIFE BIOLOGY, New York: Wiley; and J. Shaw, 1985, INTRODUCTION TO WILDLIFE MANAGEMENT, New York: McGraw-Hill, G. Caughley, 1977, ANALYSIS OF VERTEBRATE POPULATIONS, and J. Macnab, 1985, Carrying capacity and related sibboleths, Wildl. Soc. Bull. 11:397-401.) Then discuss the various ways in which the term is and has been used. Is a different interpretation useful in these situations?

2. Emphasize questions 4, 5, and 6 of **Review and Study Questions,** especially question 6 which asks students to apply the concepts of population growth to humans. Are we exceeding the carrying capacity of Earth. What is human carrying capacity, anyhow? And will our explosive population growth mean the extinction of most wild species?

# CHAPTER 13

## POPULATION REGULATION

At the outset, you should distinguish between two concepts, population regulation and population limitation. Populations are limited by any number of factors, such as disease, predation, weather, habitat quality, and the like, that set the level of equilibrium density. These factors may vary over time, establishing new levels of equilibrium density. Populations tend toward this equilibrium, labeled as K in the logistic equation, through density-dependent mortality or reproduction. Population regulation, then, involves those density-dependent processes of increasing and decreasing density-dependent mortality and reproduction that tend to bring the population back to equilibrium, set by limiting factors.

The chapter opens with a discussion of population regulation as a feedback mechanism ultimately controlled by natality and mortality. That idea is expressed in Figure 13.1. The weakness of those diagrams is their failure to incorporate environmental inputs into the system. They do not represent a systems model. I drew up such a model for the African buffalo in Figure 13.4 to show how density-dependent feedback mechanisms, driven by environmental input, rainfall, regulates the population size of African buffalo.

Density-independent influences interact with density-dependent mechanisms, and it is difficult to separate the two. Because both density-dependent and density-independent factors determine the equilibrium point, all mortalities, whatever the cause, limit the population, but only density-dependent mortality and reproduction can regulate the population. That point is elaborated by H. G. Andrewartha and L. C. Birch, 1984, THE ECOLOGICAL WEB, Chicago: University of Chicago Press.

After a consideration of intraspecific competition, the chapter turns to some of the possible mechanisms of population regulation, all of them some form of expression or outcome of intraspecific competition such as stress, dispersal, genetic change, social hierarchy, and territoriality. In a way all are interrelated. Here the material is beginning to move into behavioral ecology.

Dispersal of animals at one time had been related to crowding, a response to high densities. Ecologists now recognize that such dispersal has no regulatory impact on population. It simply reduces the population back to some current carrying capacity, K. The most important type of dispersal from both an ecological and evolutionary viewpoint is the so-called presaturation dispersal. Certain dispersal-prone individuals leave the population, settling in new habitats or joining other populations. Although such dispersal may or may not act in population regulation, it does influences gene flow, augments low populations, and extends the range of a species. Mechanisms involved in plant dispersal, brought about by wind, gravity, water and animals is discussed in an evolutionary way in Chapter 17. You will find excellent discussions of dispersal in mammals in Chepko-Sade and Halpin (1987). (See **Selected References**).

The role of behavioral interactions, involving social dominance and territoriality has been an object of study and debate for years. Examples of the role of social dominance in populations can be found in Clutton-Brock et. al., (1982) for red deer, in McCullough (1979) for white-tail deer, cited in Source Materials, and in Zimen (1978) for the wolf, cited in **Selected References.**

Territorial, because of its obvious competitive allocation of space and thus food and cover, has been viewed as a potential regulating mechanism. To regulate a population, however, territoriality must meet certain conditions: 1) a substantial portion of the population must consist of surplus animals that die, do not breed, or attempt to breed and fail; 2) a portion of the population is prevented from breeding by dominant territorial individuals; non-breeding animals are capable of breeding if dominant animal are removed; 4) breeding animals are not completely utilizing food and space. By this criteria territoriality does regulate the populations of some species some of the time.

Territoriality itself is an attribute that varies in its expression. Two papers published in mid-1985 point this out. One is J. A. Stamps and M. Buechner, 1985, "The territorial defense hypothesis and the ecology of insular vertebrates", Quarterly Review of Biology 60: 155-181. The other is J. A. Weins, J. T. Rotenberry, and B. van Horne, 1985, "Territory size variations in shrubsteppe birds," Auk, 102:500-505. The former present two hypotheses. could One is the resource hypothesis which suggests that because of resource abundance, territorial behavior is adjusted to resource densities. The defense hypothesis suggests that in addition to the influences of resources, costs of defense against both territorial intruders and contenders for vacant terries are higher. The territorial behavior exhibited results from higher available resource densities, higher defense costs, and occasionally reallocation of resources to produce more competitive young. The Wiens et. al. study in a way is a test of Huxley's elastic disk theory(J. S. Huxley, 1934, "A natural experiment on the territorial instinct," British Birds 27: 270-277., which suggests that birds have a minimum compressible size of territory and when minimal territories are attained, the habitat is saturated.

The chapter concludes with a look at space capture in plants. If one accepts the alternative definition of territoriality as the spacing out of individuals or groups more than one would expect from a random occupancy of suitable habitat, then one might view space capture as a form of territoriality in plants. They hold their space by capture of resources in a given area: light, moisture and nutrients. The idea might be stretching the point, but it should make for a good discussion and way of looking at plants from a different perspective.

## Source Materials

In addition to the **Selected References** in the text, the following provide much additional information and insights. For plants consult R. Dirzo and J. Sarukhan, 1984, PERSPECTIVES ON PLANT POPULATION ECOLOGY, Sunderland, MA: Sinauer Associates, Chapters 5, 6, and 6. H. G. Andrewartha and L. C. Birch,

1984, THE ECOLOGICAL WEB: MORE ON THE DISTRIBUTION AND ABUNDANCE OF ANIMALS, Chicago: University of Chicago Press; and by the same authors and from the same publishers , Selections from THE DISTRIBUTION AND ABUNDANCE OF ANIMALS, 1982. Social behavior and population dynamics of ungulates are discussed in T. H. Clutton-Brock, F. F. Guinness, and S. D. Albon, 1982, RED DEER: BEHAVIOR AND ECOLOGY OF TWO SEXES, Chicago: University of Chicago Press; D. R. McCullough, 1979, THE GEORGE RESERVE DEER HERD: POPULATION ECOLOGY OF A K-SELECTED SPECIES, Ann Arbor: University of Michigan Press; and M. J. Mloszewski, 1983 THE BEHAVIOR AND ECOLOGY OF THE AFRICAN BUFFALO, Cambridge, England: Cambridge University Press, as well as in A. R. E. Sinclair, 1977 and D. B. Houston, 1982. . A more general overview is C. W. Fowler and T. D. Smith, eds., 1981, DYNAMICS OF LARGE MAMMAL POPULATIONS, New York: Wiley.

## Discussion Topics

1. Do most populations reach a level at which intrinsic population regulatory mechanisms intervene, or are populations held at some lower level by extrinsic mechanisms?

2. Assume that tree cavities are very limited in a given area. Would a local population of secondary cavity-nesting birds (those that cannot excavate their own nesting places) be limited by density-dependent intraspecific competition or y density-independent environmental deficiency, the lack of cavities? What if you provided more nesting sites by erecting suitable nest boxes?

3. In our very early history, humans were subject to the same regulatory mechanisms that affect other animal population. How have humans escaped natural regulatory mechanisms? What problems do we face trying to impose our own cultural regulatory mechanisms? Do we face the risk of overshooting Earth's carrying capacity and subject humans to a massive catastrophic population crash?

# CHAPTER 14

# LIFE HISTORY PATTERNS

## Commentary

The preceding chapters considered the demographic characteristics of populations, the evolutionary outcomes of natural selection. These characteristics reflect the ways in which organisms come to terms with their environment as exhibited by differences in life history patterns.

Population growth, mortality, natality, all relate to fitness of individual organisms in a species population. The means of achieving fitness involve such economic approaches as reproductive effort, allocation of and investment of energy in young, the acquisition and defense of mates.

One aspect of reproductive effort relates to the allocation of energy to young. Fitness can be achieved only if individuals leave behind reproducing offspring. To insure that, organisms have the "choice" of investing all of their energy into one reproductive effort, semelparity, or spreading that effort over several reproductive periods, iteroparity. Organisms may elect to make minimal investments in individual young and produce many of them, or they may "elect" to produce fewer offspring with a large investment in each. Organisms may invest more energy during the preparturition or prehatching period and produce precocial young in a more advanced stage of development; or they may produce altricial young, with minimal investment in prebirth period and maximal investment in energy following birth. The number of young produced may reflect resource limitations and cost, or it may reflect some other aspect of life history such as growth pattern and developmental time .

Discussion of reproductive allocations should not be restricted to animals, however; plants, too, exhibit a wide range of reproductive efforts. Contrast reproductive efforts of annuals and biennials with perennial herbaceous species and trees. Compare the reproductive efforts of trees with wind-dispersed seeds versus those of animal-dispersed and gravity-dispersed seeds. What about the investments in asexual and sexual forms of reproduction?

Other aspects relative to reproductive effort should be explored in both plants and animals such as the relationship between size, age, and fecundity. What are the consequences to fitness if a plant or animal reproduces too early in life? How might experience, as reflected in age, improve fitness among birds and mammals? What effect might iteroparity have on fitness? Part of the answer may be found in Figure 14.1

Associated with fitness and life history patterns is the concept of r-selection and K-selection. Since it was introduced by R. H. MacArthur in 1967

("Some generalized theorems of natural selection," Proc. Natl. Acad. Sci USA 48:1893-1897) and further elaborated by R. H. MacArthur and E. O. Wilson in 1967 (THE THEORY OF ISLAND BIOGEOGRAPHY, Princeton NJ: Princeton University Press), and by MacArthur in 1972 (GEOGRAPHICAL ECOLOGY, New York: Harper and Row), the concept has undergone considerable evolution itself, carrying it far beyond the limits of the original theory. A number of life history attributes have been assigned to r-selection and K-selection (E. R. Pianka, 1970. "On r- and K-selection," Amer. Natur. 104:592-597) that have become entrenched in ecological literature and general biology texts. Organisms have even been categorized as r-species and K-species. It is an excellent example of untested ecological theory becoming ecological dogma.

Other aspects of life history parameters and fitness are mating patterns, that involve an investment of energy in the acquisition of mates which will contribute most to the individual's fitness. For both male and female, this acquisition involves both the choice and the number of mates. Mating patterns involve such problems as the advantages of monogamy over polygamy for both male and female, the quality of mates, and the like. The idea of fitness, which obviously involves mate or sexual selection, is central to evolutionary ecology. Yet like so many ideas in ecology, the concept is nebulous. Evolutionary ecology assumes that certain behaviors, certain mating patterns will result in increased fitness; otherwise why should they be selected?

But field evidence is limited. To demonstrate comparative fitness, one would have to follow the reproductive success of known individuals and known offspring over a period of time. Consult Clutton-Brock et al., (1982) (cited in **Selected References**), and Clutton-Brock, (1988 and 1991), cited in Source Materials below.

## Source Materials

Major references are cited in **Selected References** in the text. For discussions on r- and K-selection see M. S. Boyce, 1984, "Restitution of r- and K-selection as a model of density-dependent natural selection," Ann. Rev. Ecol. Syst., 15:427-447; and S. C. Sterns, 1976, "Life history tactics: A review of the ideas," Q. Rev. Biol. 51:3-47 and 1977, "The evolution of life history traits," Ann. Rev. Ecol. Syst. 8:145-172.H. Caswell, 1982, "Life history theory and the equilibrium status of populations," Amer. Natur. 120:519-529. These papers are only several of many that present different views on the subject and should not be considered in any way as the final word on the subject. Two books providing additional insights into reproductive effort and parental investment are T. H. Clutton-Brock (ed.), 1988, REPRODUCTIVE SUCCESS: STUDIES OF INDIVIDUAL VARIATION IN CONTRASTING BREEDING SYSTEMS, Chicago: University of chicago press; and T. H. Clutton-Brock, 1991, THE EVOLUTION OF PARENTAL CARE, Princeton, NJ: Princeton University Press.

## Discussion Topics

1. Consider a number of discussion questions already posed in the **Commentary**.

2. What are the pitfalls of attempting to associate life history traits with r-

and K-selection?

3. Consider the white-tailed deer of North America. By life history traits it could be considered a K-selected species, yet in parts of its range the species has reached the status of a pest, exhibiting some life history traits of an r-selection species. Discuss r-and K-selection in light of this fact.

4. Many of the concepts introduced in this chapter are applicable to human populations, but their application can be highly controversial. Consider, for example, the allocation of energy to reproduction. How is it relevant to teenage pregnancy and the age and experience of the mother? How does the allocation of energy (measured in energy, time, and money) apply to large versus small families? Does this have any effect on fitness as measured in rearing and educating the young? What is the relationship of reproductive effort and the single parent?

# CHAPTER 15

# INTERSPECIFIC COMPETITION

## Commentary

This chapter concerns one of the more controversial areas in evolutionary ecology. Leading into that topic is a review of the major relationships among individuals of population: commensalism, mutualism, predation, parasitism.

The Lotka-Volterra competition models and the competition graphs introduce the concept of interspecific competition. The models will enable the students to visualize the four major types of competition. Highly theoretical, they are based on equilibrium populations and the competitive exclusion principle.

Interspecific competition has been the cornerstone of evolutionary ecology, of major importance in the structuring of communities. Because complete competitors cannot coexist - the competitive exclusion principle - , natural selection favors individuals that are either more competitive, dominating a portion of a resource gradient, or less competitive, able to utilize resources weakly available or unavailable to other individuals. Different species occupy those segments of a resource gradient in which they can utilize the resource more efficiently than others. If competition is keen, as expressed by the amount of overlap between species on the resource gradient, species are limited to that part of the gradient which they can best exploit. That allows finer divisions of the gradient and thus an increase of number of species in the community, a phenomenon called species packing. If competition relaxes, then some of the species might expand their use of the resource gradient.

That approach, long ingrained in ecology has been challenged strongly in recent years. Interspecific competition is based on equilibrium theory and the $r$ and $K$-concept. Populations are at equilibrium level, the habitat or resource gradient is saturated, population interactions are density-dependent, and competition is a continuous selection force. The opposite view holds that habitats are not saturated, especially in fluctuating environments, populations are in nonequilibrium, competitive interactions are variable to nonexistent and felt most strongly during environmental bottlenecks such as environmentally induced food shortages.

The same side argues that field evidence for interspecific competition is rather weak. This has sent proponents of both sides seeking data to support conflicting claims. T. Schoener (1983, "Field experiments on interspecific competition," Amer. Natur. 122:240-285) reviewed over 150 field studies of competition and concluded that competition was found in 90% of the studies and 76% of their species. Competition was exhibited more frequently in experiments carried out in enclosures than in those in unenclosed areas (and thus more natural experiments). J. Connell (1983, "On the prevalence and relative importance of interspecific competition: evidence from field experiments," Amer. Natur. 122:661-696) reviewed much the same literature and concluded that competition occurred in most experiments and in somewhat more than half of the 215 species. However, most experiments did not distinguish between intraspecific and interspecific competition and where they did intraspecific competition was as strong or stronger than interspecific competition in 75% of the experiments.

The real situation, as in other ecological controversies, undoubtedly rests in some middle ground. Interspecific competition occupies one position on a gradient of interactions within and among populations. These interactions influence community structure from one extreme in which density dependence has no effect on most populations, through density-dependent regulation and delayed density-dependent responses, to interspecific competition for a limiting resource. Probably just as important as interspecific competition are predation and individual autecological responses of individuals to the environment.

The argument for autecological approach is voiced by Andrewartha and Birch, 1984. They state that competition theory is unrealistic "because it has not taken into account the multipartite structure of natural populations." (p. 209).

In spite of the controversy, students must be aware that no one, even the vocal critics of the theory of interspecific competition structures, denies that competition exists. The argument is its role in community structure. In the open environment of natural communities competition among mobile animal species may be minimal. Competition is strongest among plants. Fixed in place they have to compete with neighbors, individuals of the same or other species, for moisture, light, and nutrients. The winners are those individuals which can gain control or use the resources most efficiently. Those that lose are in effect competitively excluded, but not in quite the manner implied in the competitive exclusion principle. Competitive interactions among plants may involve allelopathy, the ecobiochemical interaction among all types of plants that may influence competitive interactions. This is as good a place as any to introduce this interaction. It will come up later.

Two new competition terms are introduced here, interference and exploitative. Exploitative competition occurs where individuals of different species use the same resources. Interference competition involves aggressive interactions between individuals of different species over access to a resource.

Competition in one way or another may influence or have influenced in the past the partitioning of resource use. At any rate, it is obvious that plants and animals do utilize different segments of the resource gradient because each is better adapted to exploit different sources or different levels. Species utilizing a common resource base in a different manner make up a guild. The term is getting wide use, especially in non-game wildlife management, although some of these guilds are rather artificial, stretching the original meaning of the term.

This leads up to the concept of the niche about which I will add little. The niche, like other ecological concepts, has been overworked, overdefined, and misused. Fortunately there is a current move back to the original definition as proposed by Charles Elton: what the organism does in its environment, its occupation. Thus habitat is the place where the organism lives; its niche is its role within that habitat. Carrying the idea beyond that, in my mind, only clouds the concept, but you may not agree. A broad niche overlap does not imply competition; in the vernacular it only means that more than one species is needed to get the job done; but if work gets scarce (i. e. if food supply is suddenly limited) then competition may result.

## Source Materials

There is an abundance of material on interspecific competition in the literature. For a start see **Selected References** in the text. The American Naturalist, Vol. 122, November, 1983, contains a group of articles on competition. It subsequently has been published (and mistitled) as a separate book, G. W. Salt, ed., 1984, ECOLOGY AND EVOLUTIONARY BIOLOGY, Chicago: University of Chicago Press. H. G. Andrewartha and L. C. Birch, 1984, present their own ideas on competition in THE ECOLOGICAL WEB, Chicago: University of Chicago Press. The role of competition in phytophagous insects is critically reviewed in D. R. Strong, J. H. Lawton, and Sir Richard Southwood, 1984, INSECTS ON PLANTS: COMMUNITY PATTERNS AND MECHANISMS, Cambridge, MA: Harvard University Press. Papers on competition appear in J. Diamond and T. Case, eds., 1985, COMMUNITY ECOLOGY, New York: Harper and Row. Papers cited in the commentary provide more insights: J. A. Wiens, 1977, "On competition and variable environments," Amer. Sci. 65:590-597; J. H. Connell, 1980, "Diversity and coevolution of predators; or, The ghost of competition past," Oikos 35:131-138; E. Connor and D. Simberloff, 1979, "The assembly of species communities: chance or competition?" Ecology 60:1132-1140; W. Arthur, 1982, "The evolutionary consequences of interspecific competition," Adv. Ecol. Res. 12:127-187; and T. W. Schoener, 1982, "The controversy over interspecific competition," Amer. Sci. 70:586-595.

## Discussion Topics

1. Interspecific competition is most obvious among cavity-nesting birds. Cavities, both natural and excavated by woodpeckers, are in short supply. Some 35 or more species in eastern North American rely on cavities for nesting sites, including chickadees, nuthatches, crested flycatchers and other as well as deer mice, squirrels, and bats. Bluebirds and tree swallows are strong competitors for nest boxes erected for the former. Assume you have erected a certain number of bluebird boxes from which tree swallows have evicted the bluebirds. What is a logical solution to the problem?

2. Assume species A is highly competitive and dominates resources at the expense of Species B. Species A is decimated by human exploitation. What effect would this have on Species B and on the success of Species A to increase following a relaxation of exploitation? This question is relevant to topics discussed in Chapter 18.

# CHAPTER 16

## PREDATION

## Commentary

Unlike interspecific competition, predation is not a highly controversial area in ecology, primarily because there is nothing subtle about it. Its outcome in individual relationships is definitive and final. What is not so obvious or fully understood is the role of predation in the regulation of prey populations. According to the Lotka-Volterra model, predation involving discrete generations results in stable equilibrium and in populations with overlapping generations, regular oscillations; but these simple models don't hold up in field situations.

Predation may be regarded as the killing of one animal for food by another, as individuals of one species eating living individuals of another, or as a process in which energy and matter flow from one species to another. The more inclusive last two definitions are used in this text. Thus it includes herbivory, carnivory, and parasitism, although nothing is said of the latter until the Chapter 17. One type of carnivory that is not included is that practiced by a small number of plant species. Carnivory in plants occurs when 1) they possess some adaptation to attract, capture, and digest prey; and 2) are able to absorb nutrients from dead animals needed to carry out their life cycle. Such plants are restricted to sunny, moist, nutrient poor habitats, particularly bogs.

The major theoretical approach to predation here is the functional and numerical responses. Because of the importance of these concepts in the study of predation, students should be thoroughly familiar with them. The graphic models in the text plot the number of prey taken against prey density. Here I add three other models plotting percent of prey taken against density of prey which may clarify the nature of the three types of functional responses for the students. For example Type III response reaches an asymptote and stays there, but when plotted using percent the curve declines after reaching an asymptote.

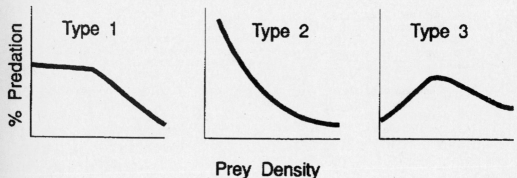

One question hangs over predation. Do predators regulate prey populations? No firm answer exists. Considering the functional and numerical responses of predation, the effects of predation on prey are usually inversely density dependent. Thus predation can destabilize a prey population, causing it to erupt or crash. If predation results in density-dependent mortality in a low prey population, predators can regulate a prey population. Predation may also regulate a prey population at low density, but fail to do so at high density. To demonstrate that predation regulates a prey population, you would have to is to remove the predators and determime if the prey population increases. If upon return of the predators the prey population fails to grow to its previous predator-free level, then you can conclude that predation regulates the population. Or you can reduce a high prey population to some low density and then determine if prey population increases. If it fails to do so, you have evidence of predation as a regulatory mechanism. For good discussions see A. Sinclair, 1989, Population regulation in animals in J. M. Cherrett, ed. ECOLOGICAL CONCEPTS, Oxford, U.K.: Blackwell Scientific Publ.

Now recognized as an important interaction between interspecific competition and predation is intraguild predation, the killing and eating of species that are use similar, and often limiting resources, and thus are potential competitors. In such predation the eater and the eaten occupy the same trophic level. Such predation accomplishes two outcomes for the predator: immediate energetic gains and elimination of an exploitative competitor. The impact on population dynamics of guild members can be complex. This topic is explored in G. A. Polis, C. A. Meyers, and R. D. Holt, 1989, "The ecology and evolution of intraguild predation: potential competitors that eat each other," Ann. Rev. Ecol. Syst. 20:297-330.

Carrying predation one step further than intraguild predation is cannibalism in which members of the same species prey on each other. Cannibalism gets considerably more attention here than in other ecology texts. It is a significant regulatory force in the populations of many species. Three papers not listed in Selected References are: M. L. Crimp, 1983, "Opportunistic cannibalism by amphibian larvae in temporary aquatic environments," Amer. Natur. 121:281-287; R. E. Buskirk, C. Frohlich, and K. G. Ross, 1984, "Natural selection of sexual cannibalism," Amer. Natur. 123:612-625; and M. Dionne, 1985, "Cannibalism, food availability, and reproduction in the mosquito fish (Gambusia affinis): a laboratory experiment," Amer. Natur. 126:16-23.

Optimal foraging theory has been an important component of predation theory. Much of the literature on optimal foraging is theoretical and mathematical. J. Krebs and R. H. McCleery provide a good overview ("Optimization in behavioral ecology") in Krebs and Davies (eds.), 1984., BEHAVIOURAL ECOLOGY. A good example of a field study of optimal foraging and a test of two models is J. C. Munger, 1984, "Optimal foraging? Patch use by horned lizards Iguanidae: Phyrnosoma," Amer. Natur. 123:654-680.

The last part of this chapter considers predator-prey systems: plant-herbivore and herbivore-carnivore relationships and their role in population dynamics. The role of herbivory in the dynamics of plant populations is reviewed by R. Dirzo (1984, "Herbivory: A phytocentric overview," Chapter 7, pp. 141-165 in Dirzo and Sarukhan; see Source Materials). He breaks herbivory into three main elements, each with its own determinants: 1) probability that plant will be found by a herbivore, influenced by the apparency of the plant and by such herbivore characteristics as search range, density, and sensory capabilities; 2) probability that the plant will be eaten by a herbivore, influenced by the plant's defenses and nutritional status and by the herbivore's feeding behavior including its nutritional requirements and detoxifying mechanisms; and 3) consequences to the plant of a herbivore feeding on it, which involves the stage of the plant's growth, the modular structure of the plant, and the quality and quantity of the tissue damaged.

The one characteristic that makes predation on plants different from predation on animals is that, except for the very vulnerable seedling stage, the plant consists of metapopulations of buds, leaves, flowers, and if woody, twigs and branches. The effect of herbivory depends upon the the proportion of leaves removed from each age class and the effect of their removal on plant fitness. In general the loss of old leaves is less damaging than the loss of young leaves; but most herbivores prefer young leaves to old. Overgrazing of a plant can adversely affect its fitness by altering its position in the community hierarchy. But in some cases herbivory may improve a plant's fitness. The response of plants to herbivory can influence community structure and composition and alter the course of succession (see Chapter 20).

Nowhere is the interaction between plants and herbivores more intense than on those plains and savannas of Africa inhabited by large grazing ungulates, who exhibit differential grazing. S. J. McNaughton (1984, 1985) has been studying the grazing ecosystem of East Africa for over a decade. He found a complex interaction between herbivores and diversity of grasses. The grazing abilities and thus nature of plant parts removed was influenced by the animals' morphology which regulated their ability to forage vegetation of different structures and ages. The grazing herbivores selected for genetic dwarfing of plants and their ability to concentrate biomass in spite of it. Studies between grases protected by exclosures and grasses exposed to grazing revealed the evolution of different genetic traits only 16 years after exclosure.

Neglected in the discussion is herbivory in marine ecosystems. We have a tendency to overlook aquatic systems. The role of plant-herbivore interactions in marine systems is reviewed in J. Lubchenco and S. D. Gaines, 1981, "A unified approach to marine plant-herbivore interactions. I. Populations and communities," Ann. Rev. Ecol. Syst. 12:405-437. Algal defenses against marine herbivore is amplified in P. D. Steinberg, 1984, "Algal chemical defense against herbivores: Allocation of phenolic compounds in the kelp Alaria marginata," Science 223:405-407.

Another evolutionary response to predation is mimicry that occurs in plants as well as animals. Our common perception of mimicry in both plants and animals extends only to the visual, but mimicry can chemical, involving smell or odors, acoustic,and tactile. For a review of mimicry see G. Pasteur, 1982, "A classificatory review of mimicry systems," Ann. Rev. Ecol. Syst. 13:169-199.

The chapter end with a brief discussion of the three-way interactions between the two systems which in northern regions which apparently is a driving force in the cyclic fluctuations in some species of northern animals. This section in effect concludes the discussion on population fluctuations in Chapter 12.

## Source Materials

Basic references are given in **Selected References** for Chapter 16. R. J. Taylor has written a good, concise, rigorous discussion of predation, PREDATION, London: Chapman and Hall. Andrewartha and Birch present a different approach in Chapter 5 in THE ECOLOGICAL WEB, already cited in commentaries of earlier chapters. D. L. Allen, 1979, provides detailed insights into predation by wolves on moose in THE WOLVES OF MINONG: THEIR VITAL ROLE IN A WILD COMMUNITY, Boston: Houghton Mifflin. Predator-prey relationship involving the wolf in Alaska are W. C. Gasaway et. al., 1983, Interrelationships of wolves, prey, and man in interior Alaska," Wildlife Monographs 84:1-50, and R. O. Peterson, J. D. Woolington, and T. N. Bailey, 1984. "Wolves of the Kenai Peninsula, Alaska," Wildlife Monographs 88:1-52. Theory of predation is well developed by M. P. Hassell, 1978, THE DYNAMICS OF ARTHROPOD PREDATION, Princeton, NJ: Princeton University Press. Optimal foraging theory is reviewed in G. H. Pyke, 1984, "Optimal foraging theory: A critical review," Ann. Rev. Ecol. Syst. 15:523-575. References in these publications will lead you into an enormous amount of literature. Long term studies of plants and grazing herbivores are described in S. J. McNaughton, 1984, "Grazing lawns: animals in herds, plant form, and coevolution," Amer. Natur. 124:863-886, and 1985, "Ecology of a grazing ecosystem: the Serengeti," Ecol. Monogr. 55:259-294. A review of chemical changes in plants responding to herbivory is reviewed in R. Karban and J. H. Myers, 1989, "Induced plant responses to herbivory," Ann. Rev. Ecol. Syst. 20:331-348. Other papers on herbivory may be found in R. Drizo and J. Sarukhan (eds.), 1984, PERSPECTIVES ON PLANT POPULATION ECOLOGY, Sunderland, MA: Sinauer Associates.

## Discussion Topics

1. One of the catch phrases in ecology has been "the prudent predator and the efficient prey." A prudent predator would exploit a prey population in such a manner that would insure a maximum yield of food and the prey population would respond by achieving maximum production. According to the prudent predator theory, the predator would remove only individuals of low reproductive value, those soon to die of other causes, and obviously the easiest to kill. Thus selection pressure from predation would favor prey individuals of high reproductive value. Based on this theory discuss:

a) Are predators prudent or are they really manipulated by the prey?

b) If predators are restricted to prey animals of low reproductive value, can they regulate a prey population?

c) Can a prey population achieve maximum production if it remains at an equilibrium level? What would result if predators removed only those individuals doomed to die of other causes? (Hint: The answer lies in Chapter 12.)

47

# CHAPTER 17

## PARASITISM AND MUTUALISM

### Commentary

The previous chapter discussed predator-prey relationships. The prey are not helpless victims of predators. Predation acts as a selective pressure on prey who in turn evolve ways of escaping predation. Predators in turn are under the selective pressures of escaping prey and evolve more efficient ways of capturing prey. Such an interaction has been termed <u>coevolution</u>. Too often, however, interrelationships between species are explained away as coevolution when no evidence exists to support it. For coevolution to occur, the two interacting species must be a distinctive selective force on each other. Many apparent coevolutionary relationships may be accidental. The behaviors of the predator and prey may have simply meshed into a paired relationship, such as the mink and muskrat as major predator and major prey. Strict pairwise coevolution is more improbable than probable. More likely is sort of a diffuse coevolution in which a general group of prey evolve responses to a general group of predators and vice versa.

I have not introduced the term coevolution until this chapter because it deals expressly with two symbiotic relationships, parasitism and mutualism, that are much stronger examples of some degree of coevolution. Parasitism is underplayed in the most ecology texts. Some fail to include the topic at all. Yet parasitism and their associated diseases can be powerful influences on populations of plants and animals. (Note that a disease is any state in the condition of a plant or animal that deviates from normal. Disease in general is cause by infections of parasites that range from viruses, bacteria, and fungi to invertebrate internal and external parasites.)

The section of parasitism deals with basics and the role of coevolution in host-parasite relationships. The classification of parasites is not by taxonomic groups but by two size classifications: microparasites and macroparasites and the functional characteristics that separate the two. Although the emphasis is on parasites of wild plants and animals, the relationship of parasites in wild populations to humans is considered, as in the example of Lyme disease. You may wish to add rabies and bubonic plague as additional examples. But providing examples of parasites is not as important as stressing modes of transmission that are basic to all kinds of parasitism. Role of parasites in regulating populations receives some attention. At this point you may bring up the role of parasitic diseases in humans over the centuries and its current impact on human populations in underdeveloped regions where medical care is almost nonexistent.

Another form of parasitism is social, both obligatory and nonobligatory. Among some species such parasitism can significantly affect their reproductive success and thus fitness. Examples include cowbird parasitism of Kirtland and other neotropical warblers and nest parasitism among certain species of waterfowl.

The second part of the chapter deals with mutualism, obligatory and nonobligatory, symbiotic and nonsymbiotic. The most widespread form of mutualism is facultative which appears to involve varying intensities of coevolutionary responses.

Because many plants depend upon animals for seed dispersal, it is only natural to assume that coevolution is involved. Plants have exploited fruit and foliage predators for seed dispersal. Herbivores, for example, consume seeds of plants along with the foliage. Because these seeds travel through the grazing herbivore's digestive tracts in tact, selection had to favor evolution of abundant small seeds with resistant seed coats. But such coevolutionary responses would be diffuse, in contrast to some of the obligate plant-pollinator systems. Plant pollination demands a mechanism for the delivery of specific pollen to a recognizable receiving point, the flower of a plant of the same species. To insure the delivery of pollen, the plant offers a reward, nectar, payed upon delivery of the pollen to the plant. Seed dispersal is different. The carrier is paid in advance with no guarantee of delivery. Seeds may be dispersed randomly over the landscape with few of them having any chance to germinate successfully. To insure some success most plants depend upon mostly upon generalists rather than specialists for dispersal.

Although we usually associate mutualism as one-to-one between species pairs, such exclusive relationships are rare in nature. Thus reciprocal interrelations tend be more general than specific. H. F. Howe (1984, ""Constraints on the evolution of mutualisms," Amer. Natur. 123:764-777) suggests four checks to mutualistic coevolution. First, diversity diffuses selection from any one source, reducing the potential for reciprocal selection among species pairs. Second, environmental disturbances over time constantly change selection regimes. Third, variations in population attributes alter the intensity of interactions between species pairs, and thus select for general rather than specific relationships. Fourth, polygenetic inheritance promotes uneven rates of evolution among mutualists. Because of these restrains, evolution of specific mutualisms would seem to occur at taxonomic levels higher than the species, rather than among species pairs.

## Source Materials

References on all aspects of coevolution discussed in this chapter are scattered through the literature. Literature citations and **Selected References** provide an entrance into the literature. Excellent sources are D. J. Futuyma and M. Slatkin (1983), Real (1983), Boucher (1985), and Howe and Westley (1988). Coevolution in plants is discussed in R. Drizo and J. Sarukhan, 1984, PERSPECTIVES ON PLANT POPULATION ECOLOGY, Sunderland, MA: Sinauer Associates. Also refer to several recent papers including: D. H. Jansen, 1984, "Dispersal of small seeds by big herbivores," Amer. Natur. 123:338-353, and N. T. Wheelwright and G. H. Orians, 1982, "Seed dispersal by animals: contrasts with pollen dispersal, problems of terminology, and constraints on coevolution," Amer. Natur. 119:402-413. Major references sources on parasitism are given in

Selected References.

## Discussion Topics

1. Many plants use foliage as bait for seed dispersal by grazing herbivores. How can this be reconciled with plants' foliage defense mechanisms?

2. Is mutualism reciprocal exploitation rather than two species acting together for mutual benefit?

3. Why are specific pair-wise interactions so rare in nature?

4. Under what conditions is a host from which a guest derives benefit likely to derive benefit from a host?

5. Question 9 in **Review and Study Questions** provides some ideas for reports and discussion. To this list you might add the spread of rabies in both Europe and North America; the relationship between spread of rabies and the "suburbanization" of raccoons; the role of mange in decimating fox population in northeastern United States.

# CHAPTER 18

## HUMAN CONTROL OF NATURAL POPULATIONS

This chapter should be of great interest for students; but to appreciate the problems discussed, they need the background information provided by the previous chapters, especially Chapter 10 through 15. In some manner, all students are at least aware of the major topics. They have experienced such insect pests as cockroaches, ants, flies, mosquitoes, mice, and rats and have undoubtedly used insecticides of some sort. They have used, observed, or know about use of herbicides on lawns and pesticides on lawns and gardens, and read about pest and weed control in croplands and forests. They buy fish in the food markets, eat shrimp in resturants, are familiar with sport hunting and fishing. They use forest products from paper, pencils and furniture to building projects. And they have read or heard about the loss of biodiversity and the growing number of species threatened with extinction, and the action being taken to save them from the temperate regions to the tropics. So you have a strong foundation upon which to discuss this chapter.

The chapter deals with three aspects of human relationships with natural populations: their control when some become pests; their exploitation, and the preservation of species. The first third of the chapter concerns pest populations. What constitutes a pest can be a controversial, as the text points out. Students should know that other types of pest control exist beyond the spray can. Familiarity with the various general types of chemical control will alert students to the ecological and public health dangers of chemical pesticides. They should recognize that persistent use of pesticides selects for resistance to pesticides. Because of the short generation time of insects and many plants, they can evolve resistance rapidly (Refer back to Chapter 2.)

Two points of emphasis are 1) a level exists in pest population where control measures are not warranted, economically or ecologically (see Figure 18.1), and 2) that control measures can produced unwanted and disastrous side effects, including massive reduction of nontarget species and predators, and interference with natural food webs and community structure.

The concept of Integrated Pest Management (IPM) should be presented. The procedure is growing and in cases involves public input into the decisions of IPM (see Decision B, Figure 18.3). Public involvement becomes important when pest control covers a large area and involves a good deal of private property from large land owners to home owners.

The second third of the chapter concerning population exploitation emphasizes the concept of sustained yield, with which students should become familiar. They should recognize the differences between sustained yield, maximum sustained yield, and optimum sustained yield; and they should recognize the the ecological and economic pitfalls of sustained yield. Sustained yield is based on single species populations. It fails to take into account interrelations among species, the impact of population reduction on food webs and n other species (see Chapter 23), and environmental vagaries. Although theoretically sustained yield should work, it rarely does because economic considerations override biological considerations. Witness, for example the exploitative methods of fishing that depletes the target resource and destroys associated species.

Problems of sustained yield also exist with populations hunted for sport. For most species we don't know if hunting mortality is additive or compensatory. Some waterfowl populations are obviously overhunted, given the disastrous decline in waterfowl nesting habitat because of wetland drainage (see Chapter 32). Other wildlife populations, particularly white-tailed deer, have grown to pest levels because of underharvesting.

Sustained yield can be most efficiently be applied to timber management. Although a great deal of basic theory and methodology exists, sustained yield is not practically that widely, even on public lands. Sustained yield forestry is becoming increasingly important and necessary. To maintain old-growth forests and large acreages of timber for wilderness, wildlife, and recreation from timber exploitation. Commercial timber lands will have to be intensively managed to insure continued production of wood products. A number of forward-looking timber companies already practice sustained yields on their own lands to ensure a continuing supply of lumber and wood products. Others, foregoing intensive timber production on their own lands, prefer to mine public owned forest lands of their timber. Many public lands, including the National Forest systems, particularly in the Pacific Northwest, Alaska, and Rocky Mountains are exploited rather than managed on a sustained yield basis.

The final part of the chapter discusses the restoration and preservation of species. Early restoration of such once endangered species as white-tailed deer and wild turkey were successful because 1) groups of highly adapted wild and not and captive-bred individuals were translocated to the wild, and 2) large expanses of habitat were available. Today for many species we face fragmented and declining habitats and an inordinate dependence on captive-reared individuals, situations which may reduce successful long-term restoration.

The major threat to world wildlife is explosive growth of human populations which results in destruction of habitats on a grand scale. Hastening the extinction of some species is poaching (made easier as animals become concentrated into habitat fragments. The wildlife trade is growing and affecting not just tropical species such as rhino but North American bears killed for gall bladders. Illegal trade extends to plants, including ginseng and cacti.

## Source Materials

Major source materials are given in **Selected References**. For a survey the contributions of wild species, both plant and animal to the resource economy of a developed economy, evaluation of genetic resources, and much more see C. Prescott-Allen and R. Prescott-Allen, 1986, THE FIRST RESOURCE: WILD SPECIES IN NORTH AMERICA ECONOMY, New Haven, Yale University Press.

NUmerous articles appear in such environmental publications such as <u>Audubon,</u> <u>International Wildlife,</u> and others. The exploitation and destruction of forests worldwide down through history and the contribution of wood to civilization is well presented in J. Perlin, 1991, A FOREST JOURNEY: THE ROLE OF WOOD IN THE DEVELOPMENT OF CIVILIZATION, Cambridge, Harvard University Press.

## Discussion Topics

1. **Review and Study Questions** 12 through 18 are major discussion topics. Here are some more:

2. What is the rationale behind the attacks within the federal government against the Endangered Species Act? Why should a significant segment of society have no concern about the future of species or the ecosystems of which they are a part?

3. When it comes to considering wildlife in any development scheme, ecologists and wildlife biologists have a difficult time assigning an economic value to it that will put wildlife values on the same level as others. Because it is difficult to assign monotary values to wildlife, wildlife is given little consideration in planning schemes. Wildlife does have monatary value usually in an indirect sort of way. Can you describe some ways in which wildlife contrbutes to the economy of a region?

## PART V

## INTRODUCTION

### The Community Defined

#### Commentary

In a way, a community is an abstraction. The term can imply an assemblage of interrelated organisms. You will see it used in the context of a grouping of a specific kinds of organisms, such as a bird community, a lizard community, or a plant community. In a broader ecological sense, the community is an assemblage of plants and animals in a given environment divorced from such functions as energy flow and nutrient cycling. No assemblage can exist under such conditions, but once those functions are included, you move from a community to an ecosystem.

## CHAPTER 19

## COMMUNITY ORGANIZATION AND STRUCTURE: SPATIAL PATTERNS

#### Commentary

Some of the more obvious characteristics of a community are growth forms of plants, stratification, and edge. Vertical stratification of terrestrial communities is rather easily observed. Observing vertical stratification in ponds and lakes requires some field work and some chemical and physical measurements. Horizontal zonation about ponds and lakes is easily observed, but some more effort may be required to measure and plot horizontal stratification in terrestrial communities. Horizontal zonation in both aquatic and terrestrial environment involves edge situations. The concept of edge goes back to Aldo Leopold, 1936, GAME MANAGEMENT, New York: Scribners, Chapter 5, pp. 124-136. Edge has since been termed patchy environment in ecological literature, but the two are one and the same. Note the shade of difference between edge and ecotone. At one time the two terms were synonymous.

Edge and ecotones differ from community gradients in which the assemblage of species changes gradually on a continuum of environmental conditions. The continuum concept of community change at one time was the focal point of controversy over the individualistic and organismal concept of the community.

You can demonstrate the continuum concept and the difficulty of mapping, naming, and describing communities by drawing a number of overlapping distribution curves similar to the ones in Figure 19.2. Use a different colored pencil for each species. Then take two rulers or straight edges to mark the boundaries of your hypothetical communities and move them along the horizontal gradient. Note how species composition and dominance changes along the gradient, depending on where you place the boundaries. Then name the community. The students will realize that naming communities depends more upon the investigator than upon nature. Discrete boundaries exist only where marked changes in environmental conditions occur.

Species dominance and species diversity, two interrelated concepts, are also abstract, especially species diversity. One is the inverse of the other. The greater a species' dominance, the lower is species diversity. The concept is important; it is useful in environmental work where ecologists use changes in species diversity to reflect changes in environmental conditions, especially aquatic pollution.

In addition to the major hypotheses of species diversity presented you should also consider the hypothesis of diversity presented by M. Huston, 1979, "A general hypothesis of species diversity", Amer. Natur. 113:81-101. Hypotheses of species diversity assume that communities exist at competitive equilibrium, and thus have to invoke competitive exclusion. However, there is considerable evidence that communities exist in a state of nonequilibrium. If a community cannot achieve equilibrium because of predation, herbivory, environmental disturbances, and the like, then the community achieves a balance between periodic reductions in populations and environmental fluctuations. Variations between these two forces can explain the patterns of species diversity without invoking other hypotheses.

How to measure species diversity has been one of ecology's problems. Is species richness a sufficient measure? What about equitability? Be sure students know the difference between the two. The most frequently used measure of species diversity is the Shannon formula, derived from information theory. It has been criticized as being non-biological. Other commonly employed methods are given in Box 19.2. These should be compared with indexes of dominance in Box 19.1. Point out that other measurements of diversity exist. A major reference is E. C. Pielou, 1975, ECOLOGICAL DIVERSITY, New York: Wiley Interscience.

Differences between community or indices of community similarity may be measured by various coefficients of community. Two of the simplest, but not necessarily the best, are given in Box 22.3. For complex calculations, including Morisita's Index and Information-Theoretic Index, see J. Brower, J. Zar, and c. von Ende, 1990, FIELD AND LABORATORY METHODS FOR GENERAL ECOLOGY, Dubuque, IA: Wm C. Brown, Publishers.

The last part of this chapter concerns insular ecology, based on island biogeography theory. The sections deal with the theory itself and some of its weakness. Students should be thoroughly familiar with this theory before proceeding into its application to real world situations. Island biography has attracted the interest of wildlife biologists and conservation biologists because they see ways in which the theory may apply to the fragmentation of major ecosystems into islands. They see in the immigration and extinction rates, species equilibrium numbers, and the size of habitat islands as a partial explanation of why so many species of animals are declining. In discussing the application of island biogeography theory, stress the major difference between oceanic islands and islands of terrestrial vegetation. The latter are surrounded by other types of vegetation, urban and suburban developments, rights-of-ways, and the like which expose the occupants of even suitable fragments habitats to increased predation and to new predators and parasites. The concept has also become applicable to the establishment, maintenance, and preservation of much of the Earth's fauna and flora. The subject of habitat fragmentation relates directly to the population genetics discussed in Chapter 2. Refer back to that chapter when discussing this topic.

## Source Materials

For a classic introduction to community ecology have student read C. E. Elton, 1927, ANIMAL ECOLOGY, out-of-print but available in libraries. For theoretical approaches to community ecology see M. Cody and J. M. Diamond. eds. 1975, ECOLOGY AND EVOLUTION OF COMMUNITIES, Cambridge, MA: Harvard University Press; D. R. Strong et. al., 1984, ECOLOGICAL COMMUNITIES, Princeton, NJ: Princeton University Press; and J. Diamond and T. J. Cade, eds. 1985, COMMUNITY ECOLOGY, New York: Harper and Row. For the best summary of island biogeography theory, see M. Williamson, 1981, ISLAND POPULATIONS, Oxford, England: Oxford University Press. For application of that theory see R. Burgess and D. Sharpe, eds., 1981, FOREST ISLAND DYNAMICS IN MAN-DOMINATED LANDSCAPES, New York: Springer-Verlag, and especially L. D. Harris, 1984, THE FRAGMENTED FOREST, Chicago: University of Chicago Press. The philosophy of species preservation and the value of biological diversity is the subject of B. G. Norton, 1986, THE PRESERVATION OF SPECIES, Princeton, NJ: Princeton University Press. Two important books dealing in part with species diversity and the problem of insularization are M. E. Soule and B. A. Wilcox eds., 1980, CONSERVATION BIOLOGY, Sunderland MA: Sinauer Associates, and O. H. Frankel and M. E. Soule, 1981, CONSERVATION AND EVOLUTION, Cambridge, England: Cambridge University Press. The plight of neotropical migrant birds both in the tropics and temperate regions is discussed in J. Terborgh, 1990, WHERE HAVE ALL THE BIRDS GONE?, Princeton, NJ: Princeton University Press. Discussing that problem and more on an international scale are P. Berthold and S. B. Terrill, 1991, "Recent advances in the study of bird migration," Ann. Rev. Ecol. Syst. 22:357-378.

## Discussion Topics

1. Concentrate on Review and Study Questions 10, 11, 12, and 13. This chapter provides an opportunity for a more detailed look at biodiversity.

# CHAPTER 20

## COMMUNITY CHANGE: TEMPORAL PATTERNS

### Commentary

The concept of succession was one of the cornerstones in the development of ecology in North America. Static for many years, interest in succession has been rekindled in recent years, because of importance in ecosystem development, evolution, and management of ecosystems. The result has been a refinement of the concept and the development of new theoretical approaches. In fact the successional studies are so dynamic that any textbook presentation of successional theory is dated at the time of publication. This commentary will update the text material to mid-1991.

The idea of succession is an old one, but it was advanced scientifically by Henry Cowles in his work on sand dune succession and Frederick Clements, in his treatise on plant succession, both of which have influenced ecology. Long-held, dogmatic ideas on succession which tended to stifle theoretical studies of it include: succession is directional, succession is autogentic brought about by the organisms themselves, and succession ends in a stable, end-point community, the climax. As discussed in the text, those ideas have been challenged and for a time split ecologists into two camps. One the one side are those who argue that succession is autogenic, that changes come within the system, producing certain emergent properties relative to maturity, nutrient cycling, energy flow, and the like. This is the ecosystem approach which is essentially a variation of Clements organismic approach. The other side argues for a reductionist approach, that changes can be explained by population dynamics and competitive interactions among plant species involved.

J. Connell and R. Slatyer (1977, citation in text) proposed three different models of succession, the facilitation, tolerance, and inhibition models. The facilitation model is autogenic; changes are brought about from within by the organisms themselves. The tolerance model involves the interaction of life history traits, especially competition. It plays a critical role in determining which species dominate during late succession. The inhibition model is purely competitive, but the ultimate winners usually are long-lived plants, even though early successional species may suppress late species for a long time. Field studies of succession discover the evidence of all three models in succession. But no matter what the approach or what the model, all emphasize some points of Clement's original theory involving nudation, migration, ecesis, competition, reaction, and stabilization.

Recent approaches to succession involving some aspects of both the ecosystem and competition approaches, in my opinion, provide best current concept of succession. One is the resource-ratio hypothesis advanced by D. Tilman, 1985, "The resource-ratio hypothesis of plant succession," Amer. Natur.

125:827-852. This hypothesis involves two components: 1) interspecific competition for resources, and 2) a long-term pattern of a supply of limiting resources, especially soil nutrients and light. Succession then results from a gradient in relative available of those resources through time. The gradient ranges from habitats with soils poor in nutrients but with a high availability of light at the soil surface to habitats with nutrient-rich soils and low availability of light. Community composition changes along that gradient as the availability of two or more limiting resources, particularly nitrogen and light, changes. Because plant species require different proportions of the limiting resources plant communities should change whenever the relative availability of those those resources change. These changes will reflect the differences in competitive abilities of plants for different ratios of soil resources and light.

In early primary succession, the colonizing species are those adapted to a low soil nutrient and high light regime. As biogeochemical processes make more soil nutrients available, plant growth increases reducing the availability of light at the soil surface. The changing ratios of soil nutrients to light leads to the replacement of one plant species by another, favoring over time plants adapted to high nutrient levels and low light availability at the soil surface. Because of the relative slowness of primary succession, succession along the gradient arrives at various plateaus of equilibrium dominated by species competitively superior at each ratio.

Secondary succession follows much the same pattern, but at a more rapid rate. Species composition and the rapidity of change depend upon the point on the gradient at which the species colonize the area. For example old field succession normally involves relatively low level of soil nutrients and high light on the soil surface. But where seed source is available and the site exhibits a point on the gradient with high soil nutrients and high light, the area may be colonized by such tree species as pines or yellow-poplar and exclude earlier successional species.

M. Huston and T. M. Smith (1987) expanded upon the idea above and proposed a nonequilibrium model of succession based on competition among individual plants to explain species replacement on a spatial and evolutionary gradient. It assumes that 1) relevant environmental resources such as light and soil nutrients change through time and among communities; and 2) the intensity of competition changes through time and among communities. The nonequilibrium hypothesis is based on three premises. 1. As plants grow, they alter the environment in such a way that the relative availability of resources changes, altering the criteria for competitive success. 2. The physiological traits of plants prevent any one species from achieving maximum competitive ability under all circumstances. 3. The interaction between 1 and 2 produces an inverse correlation between certain groups of traits such that species which are good competitors under one suite of environmental conditions are poor competitors under other conditions. Thus succession is governed by an autogenetically driven changes in resource availability through time to which plants respond relative to their adaptive abilities on a changing resource gradient.

This model is significant because it incorporates both the reductionist and ecosystem approach to succession. Succession involves both the autogenic changes proposed by the holistic approach and interspecific competition proposed by the reductionists.

As succession proceeds, community attributes change. These changes over time are discussed in detail in a classic paper: E. P. Odum, 1969, "The strategy of ecosystem development," Science 164:262-272.

Disturbance is an important component of succession and is involved in the maintenance of certain types of ecosystems. Under both the resource-ratio hypothesis and the nonequilibrium hypothesis, disturbance probably represents a shift in ratio of available resources on the gradient to which plant species respond. The role of disturbance is discussed in Chapter 21.

Aquatic succession has a few features not found in terrestrial succession. Succession is influenced by inputs of sediments from the surrounding watershed, filing in the lake basin. The accumulation of sediments increases colonization of basin by littoral vegetation. That in turn enriches the water with nutrients and organic matter, further stimulating pelagic production and further sedimentation, expanding the surface area available for colonization by macrophytes (see S. C. Carpenter, 1981, "Submerged vegetation: an internal factor in lake ecosystem succession," Amer. Natur. 118:372-383). Thus aquatic succession goes from an oligotrophic to eutrophic state and not from a eutrophic to oligotrophic state, (see Chapter 31). Succession in aquatic systems, however, is made so complex by variation and changes in watersheds that the whole process is difficult to generalize on aquatic succession.

Variations in succession are discussed in the sections on cyclic replacement and fluctuations. The section on the Climax should receive some emphasis. A mistaken idea is that climax vegetation is in a state of some flux. As senescent vegetation dies, it is replaced by new younger vegetation. This flux maintains the diversity found in climax and old growth forests. Point out that a forest of what appears to be mature trees is not necessarily a climax forest.

This chapter does not dwell on human succession on an area. That belongs more to human ecology, but human succession has a pronounced impact on natural succession. See Cronon (1983) and Irland (1982) noted below.

## Source Materials

The references complement those cited in **Suggested References.** For the role of physiological attributes of plants in succession see F. A. Bazzaz, 1979, "The physiological ecology of plant succession," Ann. Rev. Ecol. Syst 10:351-371; and F. A. Bazzaz and S. T. A. Pickett, 1980, "Physiological ecology of tropical succession: a comparative review," Ann. Rev. Ecol. Syst. 11:287-310. . For the role of nitrogen in succession see G. P. Robertson, 1982, "Factors regulating nitrification in primary and secondary succession," Ecology 63:1561-1573, and P. Vitousek, 1982, "Nutrient cycling and nutrient use efficiency," Amer. Natur. 119:553-572. For the influence of changing human society and human attitudes on the ecology of a region read W. Cronon, 1983, CHANGES IN THE LAND; INDIANS, COLONISTS, AND THE ECOLOGY OF NEW ENGLAND, New York: Hill and Wang. Picking up the story from where Cronon leaves off is L. C. Irland, 1982, WILDLANDS AND WOODLOTS: THE STORY OF NEW ENGLAND'S FORESTS, Hanover,NH: University Press of New England.

## Discussion Topics

1. Trace the succession of human communities in your local area. You will have to go back into some of the early history of the area. Is retrogression setting in? What characterized each step? What brought about the changes? What effect did human succession have on natural succession?

# CHAPTER 21

## NATURAL DISTURBANCE AND HUMAN IMPACT

### Commentary

The idea that climax vegetation arrives at some equilibrium point with the environment (climate) and remains unchanged. is dead. Vegetation is dynamic, and natural disturbances of wind, water, and fire are evolutionary components of ecosystems. This chapter deals with disturbances and their relationship to development and maintenance of ecosystems. How systems respond to disturbance is the important question and involves such concepts as stability, resilience, persistence, and resistance. In other words the balance of nature involves disturbance.

It is important to distinguish between small scale disturbance and large scale disturbance. Small scale disturbances, such as falling trees in a forest and exposure of soil in grasslands from burrowing activities of groundhogs and prairie dogs, produce gaps that permit the growth of opportunistic species and increases diversity. Large scale disturbances are much more violent: fire, windstorms, logging. They may act as a renewal agents and add diversity to landscape on much larger scale. In fact, the nature of disturbance is a matter of scale. What might be a small scale disturbance in a large area of forest could well be a large disturbance in a very small woodlot.

Many ecosystems are attuned to disturbances. Wind throw is important in maintaining diversity in tropical rain forests; drought can shift plant community composition; fire insures the continuance of fire-dominated ecosystems and fire-dependent species.

Fire is and has been both a pervasive force in shaping terrestrial ecosystems and a powerful selection force, directly on vegetation and indirectly on animals. Its importance on both has been altered greatly by human interference. At one end humans have increased the frequency and intensity of fire and at the other end have excluded fires over a long period of time. both The latter has been just as detrimental to some ecosystems as too frequent or too intense fires. Ecologists and foresters have studied and recognized the role of fire in terrestrial ecosystems, and have begun to incorporate fire in both forest and wilderness management.

Animals, too, are agents of disturbance. Their impact comes when they become too abundant and arrive at the status of pest (see Chapter 18). Insect outbreaks, such as gypsy moth, cause extensive tree mortality and influence forest growth and forest composition.

The greatest impact on forests comes from humans. As commercial forest land declines, foresters must manage forest lands more intensively, which is another form of disturbance. Regeneration and management of forests is called silviculture. Unless students have some familiarity with silviculture, it may be difficult for them to understand exactly what foresters and timber industry means when discussing issues. Methods of forest management (which often becomes mismanagement-see Chapter 18) can be controversial. Knowledge of cutting and regeneration practices would result in more ecologically and environmentally aware students.

Mining, Lewis Mumford, wrote, is warfare on the land. You need only look at the coal, gold, iron-mining regions to observe the massive effects.of mining. Major, and often irreversible changes, result. Reclamation efforts may restore grasslands, but restoration of any semblance of natural forests is remote. In the eastern forest region reclamation mostly involves conversion of forested land to grassland.

The concepts of stability, resistance, and resilience of ecosystems are basic to an understanding of how ecosystems function. Although ecosystems show some degree of stability, do not allow students to fall into the trap of believing that ecosystems will recover once the disturbance is removed. Even though aquatic systems may recover once pollution is removed, the species composition may be significantly altered. Thus disturbances can lead to a different stability domain.

## Source Materials

**Selected References** provide an entrance to the literature on disturbance. A broad survey of silviculture and silvicultural practices over a wide range of North American forests is R. M. Burns, Tech. Compiler, 1983, SILVICULTURAL SYSTEMS FOR THE MAJOR FOREST TYPES OF THE UNITED STATES, U.S. Dept. Agriculture, Agricultural Handbook No. 445. Application of silvicultural principles and the theory of island biography to the management of old-growth forest is found in L. D. Harris, 1984, THE FRAGMENTED FOREST, Chicago: University of Chicago Press. There is no one good reference book on impacts of surface mining on landscape, although the literature is extensive. An introduction is the strip mining in the Appalachians is H. M. Caudill, 1971, MY LAND IS DYING, New York: Dutton. An old but still important reference on human impact on Earth is W.Thomas, ed. 1956, MAN' ROLE IN CHANGING THE FACE OF THE EARTH, Chicago: University of Chicago Press. For a review discussion of stability see C. S. Holling, 1973, Resilience and stability of ecological systems, Annual Review of Ecology and Systematics, 4:1-23.

## Discussion Topics

1. From an ecological viewpoint has Smoky Bear been oversold? What difficulties does Smoky Bear present to modern vegetation management?

2. Report on the impacts of mountain top removal method of strip mining in the Appalachians on hydrology, forest fragments and obliteration, effects on fish and amphibians of mountain streams, acid drainage, and other effects.

# PART VI

## INTRODUCTION

### The Ecosystem Concept

#### Commentary

The introduction outlines the concept of the ecosystem. Students should be aware of the fact that the ecosystem concept is not ancient history, but only 50 years old. It had its inception in 1935 when the term was introduced by A. G. Tansley. It was 20 years later before the ecosystem concept became important in ecology.

Simple definitions of the ecosystem state that it consists of the abiotic environment plus the biotic community. Others go one step further and state that the ecosystem involves an interaction between the abiotic environment and the community. A modern, more sophisticated concept of the ecosystem views it as a cybernetic system. That view of the ecosystem is illustrated in Figure VI.1. Because generally ecosystems do not have definitive boundaries, the boundary is indicated by a dashed rather than a solid line. Inputs into the system are largely abiotic: water, carbon dioxide, nitrogen, oxygen, and most important solar radiation or energy. Outputs are similar with energy input being heat of respiration. Have students note that these abiotic inputs and outputs are not part of the ecosystem.

The ecosystem itself consists of three major subsystems or components, the producers or autotrophs, the consumers or heterotrophs, and an abiotic component involving nonliving organic matter, and soil and water nutrients. The driving forces between the compartments are consumption, deposition, decomposition, litterfall, and translocation. Note the feedback mechanisms involved.

Ecosystems are mostly autotrophic. Some may call rotting logs or caves, for example, heterotrophic ecosystems, but in reality they are embraced by the autotrophic ecosystem of which they are a part. Their energy ultimately comes from the autotrophic component of the ecosystem. The true heterotrophic ecosystem would depend upon energy fixation by chemosynthetic producers. Such ecosystems are rare. The best examples are the hydrothermal vents in the deep ocean floor of parts of the Pacific Ocean, where sulphide-metabolizing bacteria accomplish energy fixation. These vents are discussed in Chapter 34. The rest of the community, consisting of seaworms and other strange, recently discovered invertebrates, depends upon this chemosynthetic energy base.

## Source Materials

For a different discussion of the ecosystem concept see E. P. Odum, 1983, BASIC ECOLOGY, Philadelphia: Saunders. The idea that ecosystems act as cybernetic systems is not universally accepted. Some ecologists, particularly reductionists, attempt to work outside the ecosystem framework. For a review of the arguments read J. Engleberg and L. L. Boyarsky, 1979, "The noncybernetic nature of ecosystems," Amer. Natur. 114:317-324, and two replies: C. F. Jordan, 1981, "Do ecosystems exist?" Amer. Natur. 18: 284-287, and B. C. Patten and E. P. Odum, 1981, "The cybernetic nature of ecosystems," Amer. Natur. 11: 886-895.

## Discussion Topics

1. Trace the development and sophistication of the ecosystem concept since the 1950s by noting the change in the presentation of the concept in the first, second and third editions of E. P. Odum, 1953, 1976, and 1971, FUNDAMENTALS OF ECOLOGY, Philadelphia: Saunders and BASIC ECOLOGY. The first edition of FUNDAMENTALS OF ECOLOGY was the first ecology text to present the ecosystem approach in a formal manner.

# CHAPTER 22

## PRODUCTION IN ECOSYSTEMS

### Commentary

This chapter is about the flow of energy from its source, the sun, to its sink, outer space. Energy comes to Earth in two forms, heat and light. Heat energy was discussed in Chapter 4. Light energy and its role in photosynthesis was discussed in Chapter 7. This chapter looks at the processing of energy after its fixation in photosynthesis.

Before you can discuss energy flow, per se, you have to introduce four concepts: what energy is, the kinds of energy - potential and kinetic, the kinds of energetic reactions, and the First and Second Laws of Thermodynamics. If the students have had chemistry or physics, the introduction of these concepts should amount to only a quick review. But be sure students appreciate entropy.

The discussion on primary production holds to basic principles. Be sure students know the difference between production and productivity. Most of the material on primary production concerns biomass allocation and biomass distribution. Biomass allocation involves the economics of nature. How plants spend their fixed energy or biomass is important to plant fitness and survival. Note the seasonal allocations to vegetative growth, reproduction, and permanent capital in aboveground and belowground storage. The ratio of expenditures varies with the season, with environmental conditions, predation, competition and the like. This discussion of expenditures by plants can be extended to the costs of defense discussed in Chapter 16 and to life history patterns in Chapter 14. Production efficiencies are summarized in Box 22.1 rather than discussed in the text. This does not mean they can be overlooked. The terms are used throughout Part VI.

Patterns of biomass allocation show up in plant biomass distribution. Examine Figures 22.2 and 22.3 and note the similarity between biomass distribution in aquatic and terrestrial ecosystems.

Comparisons of net production among ecosystems are summarized in the text as well as in Table 22.1 and Figures 22.4, 22.5 and 22.6, which tell more than lengthy textual descriptions. Net production within ecosystems is influenced by time as reflected in the aging of plants. Older plants store much less production in biomass and considerably more in maintenance, as depicted in Figure 22.8.

Primary production supports secondary production. Secondary production begins with the conversion of primary production or plant tissue to animal tissue by the plant feeders or herbivores. How that conversion takes place is not important here. The emphasis is on what happens to assimilated energy and how it is allocated. There is a big difference between poikilotherms and homoiotherms in assimilation efficiencies and secondary production. These differences should be emphasized because of their importance of energy flow through the food chain. Note Figures 22.9 and 22.10 that point out the loss of energy within a consumer population and Table 22.2 which compares intake, respiratory costs, and production by a selected few consumer organisms.

## Source Materials

Major source materials are listed under **Selected References** in the text. Much of the material on production is summarized in the various IBP volumes, where the subject is treated in great detail. See Bliss et. al., 1981; Breymeyer and Van Dyne, 1980; Edmonds, 1982; LeCren and Lowe-McConnell, 1980; Petrusewicz, 1967; and Reichle, 1981. General material on energy is extensive. Elementary introductions are  J. J. Phillipson, 1966, ECOLOGICAL ENERGETICS, New York: St. Martin and G. T. Miller, Jr., ENERGY, KINETICS, AND LIFE, 1971, Belmont, CA: Wadsworth. Aquatic productivity is covered in W. D. Russell-Hunter, 1970, AQUATIC PRODUCTIVITY, New York, Macmillan and terrestrial productivity in National Academy of Sciences, 1975, PRODUCTIVITY OF WORLD ECOSYSTEMS, Washington, D. C.: National Academy of Science.

Energy as it relates to human affairs is discussed at length in numerous books and publications. An overall view of energy is contained in ENERGY, a Scientific American book, 1971, San Francisco: W. Freeman. A special issue of Scientific American, Vol. 148, No. 4143, 19 April, 1974, "Food and Agriculture" covers energy and human society. An excellent more recent reference is D. M. Gates, ENERGY AND ECOLOGY, 1985, Sunderland, MA: Sinauer Associates. a more advanced discussion of human society and energy flow is H. T. Odum and E. C. Odum, 1981, ENERGY BASIS FOR MAN AND NATURE, 2nd ed. New York: McGraw Hill.

## Discussion Topics

1. Why should woody plants of the tundra put more of their biomass into underground production than do woody plants in temperate regions?

2. In what way does primary production control secondary production, and conversely, how does secondary production control primary production? Refer back to Chapter 16.

3. How do humans impinge upon, impact, and redirect energy flow in ecosystems? See Odum and Odum (1981).

# CHAPTER 23

# TROPHIC STRUCTURE

## Commentary

I separated food chains and trophic levels from their usual association with chapters on energy to emphasize their importance in the relationships among organisms and in the structuring of ecosystems. The first section deals with the concept of food chains and webs and the consumer components of the food chain - the herbivores, the carnivores, the omnivores, the decomposers. But after their introduction, these traditional groups are reorganized into two main groups, biophages and saprophages. Because these two groups embrace all consumers, including decomposers, they can be grouped into their appropriate tropic levels. The problem of what to do with decomposers is solved. They are saprophages occupying different trophic levels.

This approach to food chains and trophic levels is summarized in Figure 23.3 and illustrated in more detail in Figure 23.4. It has the advantage of showing more clearly the relationship between the grazing food chain and the detrital food chain, the so called Y in energy flow through the ecosystem. The separation of energy flow into detrital and grazing pathways can mask the fact that the two are closely interrelated and that energy initially shunted down the detrital pathway can soon make its way back up into the grazing food chain. This point is clearly illustrated in Figure 23.3. On the left are the saprophages, on the right the biophages. Note how they interrelate. This concept is incorporated in the pyramid of biomass for a grassland ecosystem, Figure 23.9b. Note that the decomposers or saprophages are not set off to the side as they are are handled in other pyramids, as if they didn't quite belong to the system.

The separation of the two major food chains is distinct only at the primary producer-primary consumer level of transfer. To increase the resolution of this transfer at the autotroph-heterotroph level, E. P. Odum and L. J. Biever (1984, "Resource quality, mutualism, and energy partitioning in food chains," Amer. Natur. 124:360-376) suggest a multichannel model with six major pathways: 1) direct grazing; 2) granivory; 3) detrital particulate organic matter (POM); 4) detrital dissolved organic matter (DOM); 5) active extraction of photosynthate through mutualism and parasitism; and 6) nectarivory.

The foundation of the detrital food chain is decomposition. In the previous edition and in ECOLOGY AND FIELD BIOLOGY 4th ed., I discussed decomposition and photosynthesis together in the same chapter as complementary processes. Here I discuss decomposition as subtopic of energy flow through tropic structure, primarily to emphasize the role of the decomposer or detrital food chain as a major pathway of energy flow. I keep the discussions of the processes and function of decomposition and the role of decomposer organisms together.

Figure 23.6 points out a major fact of decomposition - that energy and nutrients contained in non-living organic matter is transferred to a host of saprophages or detrital-consuming organisms. Instead of all energy being dissipated as heat of respiration and nutrients being reduced to simplest form, they go into consumer biomass. Note that as the dead plant biomass decreases, animal biomass increases up a certain point, then declines with the reduction of plant biomass. Much of the biomass of the detrital consumers moves into the grazing food chain. Ultimately the nutrients are returned and energy is dissipated, but only only after a long, circuitous route though the ecosystem again. Nature conserves nutrients and hangs onto energy as long as possible.

Discuss the role of mineralization in the conservation of nutrients, and point out that bacteria are not only regenerators of nutrients, but also converters of nutrients into bacterial biomass that become part of the food chain, especially in aquatic ecosystems. That point will reappear in the chapter on nutrient cycling. Emphasize the different rates of decomposition and the reasons for them.

Energy flow through food chain is discussed in terms of ecological efficiencies, which are summarized in Box 23.1. Ecological efficiencies relate to production efficiencies in Box 22.1. Note the differences in assimilation and production efficiencies between poikilotherms and homeotherms as summarized in Tables 23.1 and 23.2. These efficiencies influence the rate and amount of energy flow from one trophic level to another.

A key section in this chapter is the last one, Food Webs. It introduces theoretical approaches to food chains. It considers such problems as the length of food chains, the invasion of food chains by new organisms and the deletion of others, and processes involved in the patterns of food webs which have an important role in the structure of communities. Another important question concerns the organization of food chains. Are they a random assemblage or are they not? These topics are considered here.

## Source Materials

Major references dealing with food webs and trophic level are noted in **Selected References**. A reference not included is J. E. Cohen, 1978, FOOD WEBS AND NICHE SPACE, Princeton, NJ: Princeton University Press. For more theory see Pimm 1982, Yodzis 1988, and Cohen 1989. Good sources on decomposition are Anderson and MacFayden 1976 and Swift, Heal, and Anderson 1979. Decomposition is not ignored in the IBP volumes. See in particular Bliss et. al. 1981, Breymeyer and Van Dyne 1980, and LeCren and Lowe-McConnell 1980 cited in Chapter 22.

## Discussion Topics

1. Why are some marine food chains longer than terrestrial food chains?

2. By its very nature, the pyramid of numbers cannot be inverted. Yet pyramids of numbers have been so depicted in some popular ecological writings. Such pyramids usually have as their base a single tree which in turn supports a large of insects and their predators. What is the inconsistency here? Relative to insects and the birds the tree supports, are we dealing with just a single plant? To answer, you have to consider some basic plant population ecology.

1. Why does a conspicuous leaf litter exist in an oak woods in spring while in a sugar maple and yellow-poplar stands, little leaf litter remains?

2. Compare the decomposition of pine needles with the decomposition of aspen or maple leaves, including the rate, the major detrital-feeding and decomposers organisms, and the reasons for the differences.

# CHAPTER 24

## CYCLES IN ECOSYSTEMS

### Commentary

This chapter concerns the basic principles of nutrient or biogeochemical cycles. The applied aspects of biogeochemical cycles, emphasizing human intrusions, are discussed in Chapter 25. Brief introductions to essential nutrients and to types of biogeochemical cycles are followed by discussions of the water cycle and three gaseous cycle: oxygen, carbon, and nitrogen; one sedimentary cycle, phosphorus; and one "hybrid", sulfur, with both a sedimentary and a gaseous phase. The water cycle is included here rather than earlier in Chapter 6, where you could argue justifiably that it belongs. However, because biogeochemical cycles are so closely tied to the water cycle, I have included it here. The gaseous cycles are preceded by a brief discussion of the Gaia hypothesis, about which some students may be vaguely familiar.

The carbon and nitrogen cycles are discussed in some detail. Students should be very familiar with each. The carbon cycle is tied intimately to primary production, and the carbon budgets of Earth, heavily affected by human intrusions, can influence the future course of life and functioning of global ecosystems. The discussion of carbon relates back to back to Chapter 4 and global warming. Increases in carbon dioxide concentrations in the atmosphere are illustrated in FIgure 24.5 and 24.8. The interrelationships of the various carbon pools of Earth are depicted in Figure 24.6.

The nitrogen cycle, too, is one of the most basic of all biogeochemical cycles. Students should be thoroughly familiar with the four processes of the nitrogen cycle: fixation, ammonification, nitrification, and denitrification. How humans impact the nitrogen cycle is discussed in Chapter 25.

Phosphorus is the example of a sedimentary cycle. It is a perfect one because it has no atmospheric component. I emphasize phosphorus because it is a limiting nutrient and is basic to energy flow at the cellular level. The major difficulty students will have with the phosphorus cycle is understanding its movements in aquatic ecosystems. There phosphorus is found in three forms, inorganic, organic, and particulate. Inorganic P comes from terrestrial sources, sewage disposal plants, bottom sediments, erosion, etc. The major source of organic phosphorus is excretion from planktonic organisms; it is taken up by phytoplankton and bacteria from the water. Particulate P is found in organic matter which may be ingested by zooplankton or reduced by bacteria. The aquatic phase of the phosphorus cycles relates to water pollution, discussed in CHapter 25.

Another important cycle is that of sulfur. The sulfur cycle is outlined in Figure 24.11 and the global cycle and budget in Figure 24.12. Note the importance of human intrusion in each. The outcome of that intrusion, including acid deposition, is discussed in Chapter 25.

Descriptions of nutrient cycles are followed by a discussion of nutrient flow within the ecosystem, including inputs and outputs. The example of cycling of radioactive cesium through trees provides an model of internal nutrient cycling.

External inputs of nutrients are hard to come by. Plant populations of various ecosystems have evolved mechanisms for nutrient conservation. These mechanisms vary among types of ecosystems. T. Wood, F. H. Bormann, and G. K. Voigt (1984, "Phosphorus cycling in a northern hardwood forest: biological and chemical control," Science 223:391-393) demonstrated how tightly phosphorus is cycled in a northern hardwood forest ecosystem. Phosphorus is biologically conserved by a close coupling of biological decomposition and uptake processes in surface soils. The iron- and aluminum-rich underlying B horizon of the soil functions as a massive geochemical buffer that regulates the constant but low-level losses of phosphorus to stream water. This mechanism for phosphorus retention explains why so little phosphorus is lost from the system when the forest is disturbed, in sharp contrast to the massive losses of nitrogen, calcium, and potassium to streams.

Stress the mechanisms of nutrient conservation, especially the partitioning of nutrient reserves between long-term and short-term pools. Note the role of forest arthropods in the cycling of nutrients in forest ecosystems. Also note the differences between coniferous and deciduous forest ecosystems. Conifers have evolved strong nutrient conservation measures which amount to nutrient hoarding. (See page 511, Chapter 29.).

How ecosystems operate within nutrient budgets is described briefly. The concept is introduced here, but other examples appear in Chapters 26 through 36. Compare the relatively closed cycles in natural systems with the exploitative open systems of agricultural production and timber harvest that remove nutrients from a system and transport them elsewhere. Nutrient depletion can be balanced by the addition of chemical fertilizers, but organic matter is lost. For example, nutrients removed from a midwest grain field might eventually be deposited in Russia and India or other parts of the world. Other nutrients are loss though soil erosion or leaching and become pollutants in aquatic systems.

## Source Materials

A number of general sources are listed in **Selected References**. Omitted from this list are a number of IBP volumes listed in **Selected References** for Chapter 23. Refer to this list. You will find a wealth of information and data on nutrient cycling in Bliss et. el., 1981; Breymeyer and Van Dyne, 1980; Edmonds, 1982; LeCren and Lowe-McConnell, 1980; and Reichle, 1981. Three papers on carbon cycling, all appearing in Bioscience are: J. C. G. Walker, 1984, "How life affects the atmosphere," Bioscience 34:486-491; J. Hobbie et.al., 1984, "Role of biota in global $CO_2$ balance: The controversy," Bioscience 34:492-498; and J. J. Walsh, 1984, "The role of ocean biota in accelerated ecological cycles: A temporal view," Bioscience 34:499-507. Considerable information on all phases of biogeochemical cycling can be found in F. G. Howell, J. B. Gentry, and M. H. Smith, eds. 1975, MINERAL CYCLING IN SOUTHERN ECOSYSTEMS, Springfield VA: National Technical Information Service. Students should refer to the original sources for their information on the Gaia hypothesis: J. E. Lovelock, 1979, GAIA: A NEW LOOK AT LIFE ON EARTH, Oxford, UK: Oxford University Press; and J. E. Lovelock and L. Margulis, 1974, Atmospheric homeostasis by and for the biosphere: the Gaia hypothesis, Tellus 26:1-10. See W. C. Clark and R. E. Munn eds., 1986, SUSTAINABLE DEVELOPMENT OF THE BIOSPHERE, Cambridge,, UK: Cambridge University Press and N. Myers, ed., 1984, GAIA: AN ATLAS OF PLANET MANAGEMENT, Garden City, NY: Anchor Press/ Doubleday for applications of the Gaia hypothesis to environmental problems.

## Discussion Topics

1. Few of us consider the implications of our exploitations of ecosystems. Consider the desks, chairs and other furniture, buildings, utility poles, paper. All these are wood products in one form or another; all represent accumulated biomass and nutrients that have been removed from the forest and will never be recycled in place. What is the magnitude of nutrient removal from the forest?

2. Timber management involving long rotations have minimal impact on nutrient budgets. The losses are replaced over time by inputs from the outside. But short term rotations for wood fiber and whole-tree harvesting which include branches, twigs, and even roots are another story. What would be the magnitude of losses under such management?

3. From where did the nutrients consumed by each student yesterday come? Make a list of foods, their point of origin, and nutrient contents? How are these nutrients replaced?

4. What effect does increased population growth and increased urbanization and suburbanization have on nutrient cycling? The fate of nutrients? What are the long-term implications?

5. Discuss the validity and application of the Gaia hypothesis. Why is it accepted by some in the ecologists and rejected by others?

# CHAPTER 25

## HUMAN INTRUSIONS UPON ECOLOGICAL CYCLES

### Commentary

This chapter is concerned with human impacts on biogeochemical cycles or pollution, which may be defined most simply as an overload of natural biogeochemical cycles. When the natural system is unable to process the inputs, the system breaks down with obvious results.

The water cycle reappears in this chapter which explores the ways in which we humans have affected it by drastically altering natural flows through reduction in infiltration, stream channelization and damming, and the mining and pollution of aquifers. This material also relates to Chapter 31. Refer to that chapter which considers in more depth some of the topics introduced here, if you will not cover that chapter in your course.

The topic of air pollution, the outcome of overloading sulfur and nitrogen cycles with human-induced inputs, is a major part of this chapter. Discussion of the sulfur cycle is expanded here, emphasizing its role in air pollution and its effects on vegetation and human welfare. The section on nitrogen emphasizes the contribution of nitrogen oxides to air pollution and their role in the the decline of high elevation coniferous forests, which appear to be suffering from too much of a good thing--nitrogen fertilization. A discussion of ozone naturally follows. This section should be highly relevant to students because of the news-making depletion of the ozone layer above Earth. The chemistry of that depletion is considered. Point out the ozone paradox. Above Earth its presence is essential to well-being of life on Earth; close to the surface, it is a toxic, irritating, and damaging pollutant.

Acid deposition is popularly known as acid rain, but not all acid deposition comes by the way of precipitation. A significant portion is in the gaseous and dry particulate form. Acid deposition is a controversial and politically sensitive issue. Most students are probably aware of the issue, but may have only the slightest knowledge about it. Acid deposition, its formation, and effects on vegetation, soil, aquatic ecosystems, and human infrastructures are discussed in some detail. No student should complete your course without a good understanding of the problem and issue.

The remainder of the chapter looks at three other major sources of pollution--phosphorus, heavy metals, and pesticides. Phosphorus from agricultural fields, feed lots, and sewage systems is a major cause of water pollution problems in rivers, lakes, and especially estuaries. If you will not cover a consideration of these ecosystems in your course, refer to the human impact sections in Chapters 31, 33, and 36. Lead is emphasized in the discussion of heavy metal pollution. Other excellent examples are mercury and cadium.

The input of pesticides into our ecosystems, largely from agricultural and home use, is another major environmental issue about which all students must be aware. I use DDT as the major example because its movement and effects have been studied so thoroughly. Although use of DDT is banned in the United States, it is still widely used in Third World countries, so we are still exposed to its environmental effects. Emphasize the amount of pesticide use; in particular point out that home owners and gardeners are the major purchasers and users of pesticides. Both groups, and that embraces most of us, can have the capability of significantly reducing pesticide use. Emphasize how pesticides and herbicides affect community structure and composition, interfere in cycling of nutrients, and human health.

## Source Materials

Major sources are given in **Selected References** at the end of the chapter, but literature on the subject is vast. Students certainly should be familiar with Rachael Carson's classic SILENT SPRING, available in various hardback and paperback editions. A overview of the nature and politics of acid is reviewed in C. Park, 1989, ACID RAIN: RHETORIC AND REALITY, New York: Chapman and Hall. Analyzing the water situation in western United States is M. Reisner and S.F. Bates, 1989, OVERTAPPED OASIS: REFORM OR REVOLUTION FOR WESTERN WATER, Washington, DC: Island Press. Methods of managing gardens and lawns without a heavy dependence on pesticides are provided in A. Carr et. al, 1991, CHEMICAL-FREE YARD AND GARDEN, Emmanus PA: Rodale Press.

## Discussion Topics

1. Find out exactly what chemicals are involved in local lawn treatment programs. You may experience difficulty in getting straight answers for this question.

2. Set up a simple monitoring device to measure the pH of local rainfall over a period of several months.

3. A report shows that 90 percent of samples of swordfish found in fish markets had detectable levels of mercury ranging from 0.46 to 2.4 parts per million. Informal standards for methylmercury is 1 part per million. Trace the sources and pathways of mercury from industrial and household wastes and burning fuels to its accumulation in large marine fish. What is the effet of mercury on humans?

4. Comment on the statement "Short supplies of clean water could rival expensive oil as one the nation's most serious concerns." What is the basis for this statement?

# Part VII

# DIVERSITY OF ECOSYSTEMS

PART VII

INTRODUCTION

Patterns of Terrestrial Ecosystems

## Commentary

Although most ecology texts consider the major ecosystems, they do so in a very cursory manner with little more than a brief description of each. I have always maintained that students of ecology should should have a deeper knowledge of major ecosystems, especially their structure and function. Current environmental issues of great concern involve ecosystems: wetlands, old-growth forests, overgrazing of grasslands, fragmentation of forests, ocean pollution, overdevelopment of seashores, to mention a few. An understanding of and participation in the issues requires more than a simple recognition of what these ecosystems are. Students also need to be aware of the effects humans have on these systems. Achieving that goal is the objective of this final part of the text. It is, in a way, the application of some of the principles presented earlier in the text to particular ecosystems.

The introduction examines terrestrial communities relative to their distribution over Earth. The emphasis on the various approaches employed to associate the distribution of terrestrial ecosystems with continents and climates, beginning with biogeographical realms, associated most strongly with animal distribution, followed by life zones, biomes, and ending with the Holdridge life zone system.

## Source Materials

Both C. B. Cox and P. D. Moore, 1985 BIOGEOGRAPHY: AN ECOLOGICAL AND EVOLUTIONARY APPROACH, 4th ed. Boston: Blackwell Scientific. and J. H. Brown and A. C. Gibson, 1983, BIOGEOGRAPHY, St. Louis: C. V. Mosby provide good introductions to the subject. For the biome concept see F. E. Clements and V. E. Shelford, 1939, BIO-ECOLOGY, New York: John Wiley; for life zones, C. H. Merriam, 1898, LIFE ZONES AND CROP ZONES OF THE UNITED STATES, Bull. U. S. Bureau of Biology Survey, available in some libraries. This publication is the original source of the concept. For the original description of the Holdridge life zone system see L. R. Holdridge et. al., 1971, FOREST ENVIRONMENTS IN TROPICAL LIFE ZONES: A PILOT STUDY, New York, Pergamon.

## Discussion Topics

1. Compare the usefulness of the various approaches to the classification of terrestrial ecosystems. What are the advantages and disadvantages of each?

# CHAPTER 26

## GRASSLAND AND SAVANNA

### Commentary

The first to be considered is the grassland ecosystem. Over many parts of eastern North America and in Britain and Europe, tame grasslands are the major grassland ecosystems. Because they are accessible and relatively simple, compared to forests and shrublands, grasslands present excellent sites for field studies in an ecology course. Because lawns, hayfields, and pastures, particularly the first, are grassland ecosystems most familiar to students, you may suggest that students read or incorporate into your lecture the following paper: J. H. Falk, 1976, "Energetics of a suburban lawn ecosystem," Ecology 57: 141-150. This paper should stimulate ideas for a laboratory study of the most accessible of ecosystems, the lawn. Lawns offer many possibilities for study including not only energetics, but species diversity, plant competition, herbivory above and below ground, predation, and other relationships.

Natural grasslands worldwide have severely impacted by human activity. They either have been converted to cropland or overgrazed by domestic livestock. The distinctive forms of grassland wildlife from American bison to African antelope have been reduced to remnant populations, their places taken by domestic grazing animals. Species of wildlife, especially birds, inhabiting tame grasslands, especially in eastern North America, are also declining because of changing land management practices, land abandonment, and conversion to suburban developments. A review of the status of world grasslands and their wildlife will bring the problem sharply into focus for students.

Grasslands show a strong relationship to rainfall and grazing pressure. A discussion of grassland function and structure and the productivity of grassland vegetation as they relate to grazing pressure offer an opportunity for further discussion of coevolution. You can relate function and productivity to the response of lawns to mowing, a form of simulated grazing.

Tropical savannas and woodlands cover much of the Southern Hemisphere, but they are familiar to students of the Northern Hemisphere only as they se them on television. Savannas are the homes of much of our spectacular wildlife. Like grasslands, savannas have been severely impacted by human activity, ranging from destruction of savanna woody vegetation for firewood and overgrazing to conversion to cropland and plantation forestry at the expense of native grazing and browsing ungulates.

Minimal information on structure and function of savannas is available, but literature is accumulating. Students may wish to explore that literature, much of it scattered throughout many international journals. See **Selected References** and **Source Materials** for an entrance into the literature.

## Source Materials

Basic references are given in **Selected References.** For general reading on grassland ecosystems, students should be directed to more popular reading which are both informative and entertaining. A detailed account of the original prairie is contained in J. A. Weaver, 1954, THE NORTH AMERICAN PRAIRIE, Lincoln: University of Nebraska Press, and in J. A. Weaver and F. W. Albertson, 1956, GRASSLANDS OF THE GREAT PLAINS, Lincoln, Nebraska: Johnsen. Those two books concentrate in the vegetation. For more sweeping accounts see D.W. Allen, 1969, THE LIFE OF PRAIRIES AND PLAINS, New York, McGraw-Hill, and Reichman (1987). Tame grasslands with emphasis on Great Britain are considered in E. Duffey et al. (1974). Two comprehensive books on grasslands are R. T. Coupland, ed., 1991, NATURAL GRASSLANDS, and R.W. Snaydon, ed., 1987, MANAGED GRASSLANDS, New York: Elsevier Science Publishing. For savannas, see also a book not included in **Suggested References,** F. J. Huntley and B. W. Walker, 1982, ECOLOGY OF TROPICAL SAVANNAS, New York: Springer Verlag. For the development of agriculture in savanna regions, see J. M. Kowal and A. H. Kassam, 1978, AGRICULTURAL ECOLOGY OF SAVANNA, Oxford, UK: Oxford University Press.

## Discussion Topics

1. Compare the degradation of grasslands in Africa and Australia with those in North America. Is there a common pattern? Why or why not?

2. What are come ecological indicators of overgrazed grasslands? (Ecologically aware students should be able to pick some of these indicators by observing grazed grasslands along highways.)

3. Expand upon the discussion of the coevolutionary relationship between grazing ungulates and grasses. (For a start see S. J. McNaughton, 1984, Grazing lawns, animals in herds, plant form, and coevolution, <u>American Naturalist</u> 124:863-886.

4. Where has agricultural development of African savannas been most intense? What crops are involved? How does agricultural development impact savanna wildlfe?

# CHAPTER 27

## SHRUBLAND AND DESERT

### Commentary

Shrublands are poorly treated in most ecology texts, if at all. Only in recent years have they been receiving the attention they deserve. However, most studies relate to semi-arid mediterranean-type ecosystems, the dominant form of climax shrubland. These shrublands occur in the Mediterranean region, southern California and Arizona, Chile, South Africa, and southwest Australia where shrubland is dominated by Eucalyptus. In North America the northern desert scrub, dominated by sagebrush, is a widespread formation. Although shrubland, it is classified as cool desert. High elevations around the world are dominated by heathlands. Seral shrublands, mostly a stage on the way to forest, are important to many species of wildlife. Because of the data available, the discussion on structure and function relates to mediterranean systems. They provide a good example of adaptations to wet winters and dry summers, and to fire.

Shrublands, climax or seral, have been devastated by human activity. The grandeur of ancient Greece and the glory of ancient Rome were build upon the exploitation of the mediterranean-type vegetation about the Mediterranean Sea and has left that region with the impoverished soil and degraded vegetation of today. Southern California is rapidly following the same pathway. Seral shrubland are regarded by tax assessors and developers as wastelands prime for development. Yet many species of wildlife depend upon those ephemeral shrublands as habitat. The decline in shrubland birds can be attributed to either the destruction of those shrublands or their passage into forest vegetation.

Deserts occupy distinct belts around the world. Relate the location of deserts to climatic patterns. Distinguish between a cold desert dominated by sagebrush and similar species, and a hot desert. Although our basic example of cold deserts is sagebrush of North America, similar cold deserts occur in central Asia. Students should be aware of the fact that not all deserts are hot. In fact, the tundra, discussed in Chapter 28, is also a desert of a different sort.

The functioning of desert ecosystems differs considerably from that of other ecosystems. Production comes in pulses and depends upon the proportion of available water used and the efficiency of plants using it. Primary production is low. Compare the productivity among deserts and compare deserts with other ecosystems. Point out the limited role of detrital food chains in desert ecosystems. Adaptations to arid environment, especially by animals, is not stressed in this section. If you do not emphasize physiological ecology, then adaptations to arid environments discussed in Chapter 6 can be introduced here.

In spite of their aridity, deserts have been severely impacted by human occupancy and development. The role of humans in desert regions, the adaptation of some people to the desert environment, and the effects when that ecological balance is broken by invasions of maladapted human settlements makes good subjects for student reports. A point also to be stressed is the role of humans in desert expansion into regions bordering them, a process termed desertification.

## Source Materials

Major references for this chapter are given in **Selected References**. See, in particular, Castri et al (1981), Evenardie et al. (1985-1986), and Specht (1979, 1981). The role of fire in mediterranean-type ecosystems is well covered in H. A. Mooney and C. E. Conrad, eds., 1977, PROCEEDINGS: SYMPOSIUM ON THE ENVIRONMENTAL CONSEQUENCES OF FIRE AND FIRE MANAGEMENT IN MEDITERRANEAN ECOSYSTEMS, Gen. Tech. Rept. WO-3, Washington DC: USDA Forest Service. Not listed in the text is N. West (ed.) 1983, TEMPERATE DESERTS AND SEMI-DESERTS, New York: Elsevier. A general book on desert life is J. L. Cloudsley-Thompson, 1954, BIOLOGY OF THE DESERTS, Hafner: New York. A general reference on North American deserts is E. C. Jaeger, 1957, NORTH AMERICAN DESERTS, Stanford, CA: Stanford University Press. This book is old, thus providing a description of the southwestern desert as it used to be. For a survey of animal life of the deserts of the world read F. H. Wagner, 1980, WILDLIFE OF THE DESERTS, New York: Harry N. Abrams. A classic work on the desert that should be read for description and philosophy is J. W. Krutch, 1952, THE DESERT YEAR, New York: Slone, but available in paperback editions. A new book in somewhat the same vein by an animal behaviorist , J. Alcock, is SONORAN DESERT SPRING, 1985, Chicago: Chicago University Press. A good starting point for a report on desertification in the Sahel of west Africa is National Research Council, 1983, ENVIRONMENTAL CHANGE IN THE WEST AFRICAN SAHEL, Washington DC: National Research Council.

## Discussion Questions

1. **Review and Study Questions** 5, 9, 10, and 11 will engage students with some of the major problems of shrubland and deserts. Here are some additional ones.

2. Pressures of population growth push development into fire-prone ecosystems. Such human intrusions and unregulated expansion result in fragmented land ownership, accumulation of highly inflammable fuels, increased probability of ignition, and thus the frequency of devastating fires. What are some solutions? What precautions can homeowners take? A good place to start is Mooney and Conrad (1977) and J.B. Davis and R. E. Martin, tech. coord., 1987, PROCEEDINGS OF THE SYMPOSIUM ON WILDLAND FIRE 2000, Gen. Tech. Rep. PSW-101, Berkeley, CA: Pacific Southwest Forest and Range Experiemnt Station, Forest Service, U. S. Department of Agriculture.

3. What is the status of shrublands in your area? What is their fate? What is the population status of shrubland wildlife, especially species of seral shrublands?

# CHAPTER 28

## TUNDRA AND TAIGA

### Commentary

Because protecting the fragile Arctic tundra is at the heart of the oil exploitation controversy in Alaska, particularly the Arctic National Wildlife Refuge, this chapter should hold special interest to students. The more knowledgeable they are of the nature, structure, and function of the tundra, the better they will understand the issue.

The tundra is an arctic desert. Lacking in rainfall, it maintains its wetness because of permafrost that impedes drainage and cold temperatures that inhibit evaporation. Shaped by climatic conditions, the tundra landscape features frost hummocks, stone polygons, and solifluction terraces. Both plant and animal life are well adapted to life in cold environment. This section emphasizes adaptation of plant life to the tundra environment.

Stress the differences and similarities of the arctic and alpine tundra. Emphasize the fact that alpine tundras are not restricted to the Rocky Mountains and the Alps. They occur in all of the very high mountains of the world. Because the arctic tundra has been studied the most intensively, the discussion of function emphasizes arctic tundra. Note how the severe physical environment influences both structure--low growth and bulk of biomass in the roots--and function, especially nutrient cycling, restricted to short growing season. rapid recycling, and animal input.

The fragility of the tundra ecosystem makes it very vulnerable to human disturbance. The traditional inhabitants of the tundra, the Eskimos, northern Indians and Lapps, live, or once lived, in harmony with the environment. However, the intrusions of modern technology has changed all that by breaking down the close ecological relationships. Further the exploitation of the Arctic for oil and other minerals threaten the tundra and its animal life. The problems of the tundra provide the opportunity to relate ecological concepts to environmental problem.

Below the tundra lies the taiga, a circumpolar belt of coniferous forests that make up Earth's largest plant formation. Unlike the tundra, the taiga, in spite of its immensity, is one of the least familiar to students. They know it is there, but little else. Points of emphasis should be the four major vegetation zones within the taiga, the role of permafrost in the structure of the taiga, the adaptations of plants to the cold environment, and the effect of low availability of nitrogen to forest trees.

This inherent infertility can have a major impact on the future of the taiga. The taiga is a major source of pulpwood. Much of the vast forest is mined rather than managed for timber production. Without careful management, the taiga forest may revert to muskeg and not be replaced by new stands of spruce.

In addition, the taiga is a rich source of minerals, including nickel and iron ore, the exploitation of which can cause irreparable damage to the taiga ecosystem. Massive hydroelectric schemes threaten to flood vast areas of taiga, diminishing its wildlife resources and destroying Indian cultures. The problems of the taiga again present students with the opportunity to combine ecological knowledge with social and political problems impinging on the environment.

## Source Materials

Basic references are listed at the end of Chapter 28. Two books relating to the tundra ecosystem not listed are L. L. Tieszen, ed., 1978, VEGETATION AND PRODUCTION ECOLOGY OF AN ALASKAN ARCTIC TUNDRA, New York: Springer-Verlag; and W. Tranquillini, 1979, PHYSIOLOGICAL ECOLOGY OF THE ALPINE TIMBERLINE, New York: Springer-Verlag, dealing with the krummholz. A fine general introduction to the Arctic is the beautifully written and beautifully illustrated THE ARCTIC by Fred Bruemmer, 1974, New York: Quadrangle. Animal life in the arctic is covered in B. Stonehouse, 1971, ANIMALS OF THE ARCTIC: THE ECOLOGY OF THE FAR NORTH, New York: Holt, Rinehart, and Winston. Description of the tundra in Europe is covered in W. H. Pearsall, 1960, MOUNTAINS AND MOORLANDS, London: Collins. Botany and plant ecology of the alpine tundras of North America are well described by A. H. Zwinger and B. E. Willard, 1972, LAND ABOVE THE TREES, New York: Harper and Row. A lyrical and appealing description of arctic tundra is B. Lopez, 1986, ARCTIC DREAMS, New York: Scribners. The taiga does not receive such treatment; best overall references are Larsen (1980, 1989). Implications of hydroelectric development in the taiga of northeastern Canada are described by H. Thurston, 1991, Power in a land of remembrance, Audubon 93(6):52-59.

## Discussion Topics

1. **Review and Study Questions** 10 and 11 suggest highly relevant topics for for discussion and reports, especially question 11.

2. Exactly what are the impacts of such human activity as hiking and sking have on the alpione tundra? Should tourisn to alpine tundras be restricted?

# CHAPTER 29

## TEMPERATE FORESTS

### Commentary

Of all terrestrial ecosystems, forest have received the most attention because of their economic importance. Careful management of forest ecosystems depends upon an understanding of their structure and function. For that reason studies of forest ecosystems, originally the domain of plant ecologists, have been dominated by forest scientists, and much of the literature now appears in forestry related journals. These studies, however, do not preclude that the findings are applied to on-the-ground forestry. In fact in too many cases, they are not, as evidenced by the management of old-growth forests, extensive clear-cutting, short-term, nutrient-depleting rotation forestry, and the like.

Structurally forests are the most diverse and complex of all terrestrial ecosystems. To emphasize this diversity I divided the temperate forest into two broad categories, coniferous and broadleaf deciduous and discuss them separately. In each section I provide a very broad classification of the major types. There is some overlap here with the discussion of boreal coniferous forests. The montane coniferous forests are sometimes included in the boreal coniferous forests as are the outliers of spruce and fir in the high Appalachians. And components of the boreal forest are seral deciduous forests of aspen and birch.

The sections on structure in both coniferous and broadleaf deciduous emphasize growth forms and vertical stratification of biomass that in turn influence the microclimate within the forest and stratification and diversity of animal life. Note the differences in growth forms of the several types of conifers and their influence on coniferous forest stratification. Stratification in broadleaf deciduous forests is strongly influenced by the age structure of the forest, competitive interaction among trees, and disturbance regimes. Even-aged stands which result from such major large disturbances as fire and clearcutting, go through a growth period, called "pole-stage" by foresters, in which stratification is minimal and dominated by one layer, the canopy. In contrast old-growth and uneven aged stands have much more structural diversity.

The sections on function emphasize nutrient cycling, with particular reference to the nitrogen cycle. For the coniferous forests I contrast nutrient cycling in young stand versus an old-growth stand of Douglas-fir. Of significance is the role of the canopy microcommunity in nutrient cycling in old-growth forests, a mechanism destroyed when old-growth forests are clearcut. Also emphasized is the role of dead wood in forest ecosystem. Although introduced in the section on coniferous forests, dead wood plays a similar role in the broadleaf forests. Differences in function between coniferous forests are pointed out in Tables 29.1, 29. 2 and 29.3. Nutrient cycling in coniferous and deciduous forests are compared at the end of the section on Function in broadleaf forests and in Figure 29.11.

Human impacts on temperate forests have been intense. The forests of today, whether in Europe or North America. Over the centuries, most of the forested landscape was cleared, tilled, abandoned, regrown, cut, burned over, and regrown and being cut again. Much of second forest in North America arose on abandoned agricultural land, and its composition did not match that of the original. The second forest has been replaced by a different set of species, and the third forest is subject to heavy management and artificial regeneration. About 9 percent of all commercial forest lands in the United States are artificially regenerated. Most European forests long ago ceased to be natural. Further large expanses of forests are being fragmented into subdivisions and bisected by roads, that will forever change the nature and character of the eastern deciduous forest and its animal life. The impact of these changes have not yet been realized by most of us. Some of the most intense impacts are taking place in the Pacific Northwest where once vast tracts of forest 500 to 1000 years old have reduced to about 8 percent of original expanse. Such cutting follows the pattern of logging plunder that deforested North America. Old growth forest are being replaced by short-rotation plantation forest that can never support the original diversity of life or highly evolved complex ecosystems of old-growth stands. Because the issues of clear-cutting, the conflict of timber cutting with recreation, wildlife habitat, and aesthetics are intense, featured in news articles, magazine and television specials, this chapter requires a special emphasis, if students are to understand basics of the issues. REfer back to Chapter 18, pp. 285-286 and Chapter 21, pp. 349-350.

## Source Materials

**Selected References** provide an entrance into the vast amount of literature of forests and forestry. Application of silvicultural principles and the theory of island biogeography to the maintenance of old-growth forest is L. D. Harris 1984, THE FRAGMENTED FOREST, Chicago: University of Chicago Press. A basic reference on old-growth forest of the Pacific Northwest is J. Franklin et. al., 1981, ECOLOGICAL CHARACTERISTICS OF OLD-GROWTH DOUGLAS-FIR FORESTS, U. S. D. A. Forest Service General Technical Report PNW-118. Also see E. A. Norse , 1989, ANCIENT FOREST OF THE PACIFIC NORTHWEST, Washington, DC: Island Press. for an excellent overall account. W. Cronon, 1983, CHANGES IN THE LAND: INDIANS, COLONISTS, AND THE ECOLOGY OF NEW ENGLAND, New York: Hill and Lang, chronicles the relationship of humans and the forest in New England since presettlement, the destructive effects of the English on the New England forests and the ecological impacts of that deforestation. Providing a world-wide survey with excellent chapters on North America is J. Perlin, 1991, FOREST JOURNEY, Cambridge, MA: Harvard University Press.

## Discussion Topics

1. Explore the effects of changing land use policy and land ownership in United States on forest ecosystems and its animal life, especially interior species. loss of potential timber producing forests, and impact on public forest lands.

2. Characterize the major ecological differences and advantages and disadvantages between even-aged management involving clear-cutting and selection harvesting of forest. This topic may be a difficult for some to probe, but it is at the heart of the debate over managing forests.

3. The following appeared in a book by Franz Heske in 1938 (GERMAN FORESTRY, New Haven: Yale University Press) writing relative to German experience of one-stand artificial forests over 100 years:

> German experience confirms the biologic fact that the forest is a complicated community of living beings, in which each tree species is merely a member, no more and no less important for the health of the whole than the other members. A single species may not be used with impunity in plantations where it is entirely isolated from its natural organic complex. The foundation and elements of practical silviculture are not the individual species of trees, but the natural communities of which these economically desirable trees are a part. The growing of less valuable, but biologically important, species in mixture with those of high economic value is equivalent to paying an insurance premium against later losses.

This statement was written before ecology emerged as a science, yet it reads as if it were written today. If some foresters already recognized the value of natural forests and the great biological disadvantages of depending on artificially regenerated monocultural "forests", why do many foresters and the timber industry fail to heed ecological truths? Is this deficiency embedded in training, philosophy, economics, or what?

4. Was the real reason for the deforestation of New England to provide lumber or firewood? Compare the demand for and use of wood in New England with the demand for wood in England in the late seventeenth and eighteenth centuries? What impact did forestry activities in both have on the forests of today?

# CHAPTER 30

## TROPICAL FORESTS

### Commentary

It would be unusual for any student to unaware of tropical deforestation, especially the burning of the Amazon. They are probably much more conversant about the problems of tropical forests in Brazil than they are of the deforestation of the Pacific Northwest. Students, however, may not fully understand the reasons behind or the extent of tropical deforestation. In 1989 more than 55,000 square miles of tropical forests were lost around the world. Much of this is in southeast Asia. Tropical forest are gone in Thailand; they are reduced to a fraction of its original in peninsular Malaysia; Because the Malaysian states of Sabah depends sole on timber for its income, the tropical forest is being liquidated and will gone within 10 years. The situation in Sarawak is not much better. The forest of many of the tropical Pacific Islands are also being assualted. Logging is one of the primary reasons for tropical forest destruction. Many other millions of acres are cleared for agriculture--cattle ranching in South America, rubber and oil palm plantations in southeast Asia, slash and burn agriculture in Central and South America and the Philip;ines. Tropical forests are drowned beneath dams and overturned in mining operations.

The response of tropical forests to disturbance varies with the the type of disturbance. Logging, per se, does not destroy a tropical forest. Because most species of trees require light to regenerate, the removal of the canopy stimulates new growth. The problem of logging is its scale. The diversity of the tropical forest and regeneration of its tree species is maintained by creation of big gaps through natural disturbance. Logging covers a much larger area. Thus the removal of widely scattered emergent trees and the species of the upper canopy eliminates flower and fruit production for a long period of time upon which a wide array of fruit-eating birds, mammals, and pollinating animals depend.

Patches of forest completely cleared for swidden agriculture and then abandoned are colonized by fast growing pioneering species. These forests are termed secondary. In places the growth may extremely tangled and dense, giving rise to the the popular concept of the jungle. Huge acreages of land cleared and burned for cattle ranching may never return to a tropical forest, or at best a degraded form.

A major tragedy of the logging and clearing of the tropical forests is the fate of the native peoples, whose rights are rarely respected. Living for thousands of years in the tropical forest, these people are dependent upon the rain forest for their livelihood and way of life. Governments usually take the view that native peoples should be removed and placed in Western-style villages, divorced from the tropical forest environment they know. Their loss is

88

our loss for these people alone know the thousands of species of life in the tropical forest, their uses and medicinal values. As these people disappear, that knowledge transmitted from generation to generation also disappears, along with the rich genetic diversity the tropical forest holds.

## Source Material

Major references are listed in **Suggested References.** However, interest and concern over tropical deforestation and the loss of biodiversity is spawning additional titles. One is K. Miller and Tangley, 1991, TREES OF LIFE: SAVING TROPICAL FORESTS AND THEIR BIOLOGICAL WEALTH, Boston: Beacon Press, is a primer on tropical forests, including up-to-date information on various aspects including rates of deforestation, history and uses of tropical forests, economic pressures, and sets of practical recommendations to solve. J. C. Kricher, 1989, A NEOTROPICAL COMPANION, Princeton, NJ: Princeton University Press, is an excellent introduction to plants, animals, and ecosystems of the New World Tropics. J. Gradwell and R. Greenberg, 1988, SAVING THE TROPICAL FORESTS, Washington, DC: Island Press describes factors contributing to worldwide deforestation, along with case studies. R. Goodland, ed., 1990 RACE TO SAVE THE TROPICS: ECOLOGY AND ECONOMICS FOR A SUSTAINABLE FUTURE, Washington, DC: Island Press explores application of ecology to national development and conservation plans. S. Hecht and A. Cockburn, 1989, THE FATE OF THE FOREST: DEVELOPERS, DESTROYERS, AND DEFENDERS OF THE AMAZON, London: Verso. For a good detailed discussion of tropical forest destruction in Sarawak and Saba see S. Sesser, Logging the Rain Forest, New Yorker, May 27, 1991. D. J. Mabberley, 1991, TROPICAL RAIN FOREST ECOLOGY, 2nd ed., New York:Chapman and HAll is an excellent primer on the subject.For a more popular account of tropical forests and their destruction read C. Caufield, 1986, IN THE RAIN FOREST, Chicago: Chicago University Press.

## Discussion Topics

1. Discuss the silvicultural methods developed for the sustainability of the tropical rain forest. Why are these practices not widely employed? See Collins (1990), Miller and Tangley (1991), Whitmore (1984).

2 What resources come from the tropical forests of the world? Give particular emphasis to medicinal and food plants. (For a start see Myers 1983, cited in **selected References.)**

3. What countries are the major consumers of tropical timbers? Why?

4. What do we lose when indigenous peoples are become extinct or removed from to forest and established in resettlement villages. (For a start see E. Linden, Lost Tribes, Lost Knowledge, Time, September 23, 1991.)

5. Chronicle the deforestation of the Hawaiian Islands and its impact on the indigenous fauna. Is development and the tourist industry contributing to further deforestation? Are there any efforts to save the remnant tropical forests?

# CHAPTER 31

## LAKES AND PONDS

### Commentary

There are two types of freshwater ecosystems: running water or lotic and still water or lentic. If you consider all aquatic ecosystems as a continuum, the extreme lotic system is a fast mountain brook and the extreme lentic system is a lake or pond with minimal inflow or outflow of water. The latter condition could develop into wetlands of various types, so freshwater ecosystems also include swamps, marshes, and peatlands, although they are half-way worlds between aquatic and terrestrial environments.

Discussion of lentic systems parallels that of lotic systems with emphasis on physical characteristics and zonation, both vertical and horizontal. Horizontal zonation embraces the littoral zones from submerged macrophytes to shallow water emergents. The development of the littoral zone is influenced by water depth. Shallow basins and shore lines have well-developed littoral vegetation. Emergent vegetation drops off sharply as depth of water increases. THus steep-sided lakes and ponds have a poorly developed littoral. Lentic systems with well developed littoral vegetation support a higher diversity of invertebrates, amphibians, and fish. Many of the latter frequent weed beds to feed on the abundance of invertebrate life.

Lakes are strongly autotrophic and nutrient cycling differs markedly from that of lotic systems. Cycling depends on spring and fall overturns, phytoplankton and zooplankton interactions, contributions from littoral macrophytes, and nutrients inputs from the surrounding watershed. Because lakes and ponds serve as catch basins for the surrounding watershed, they are strongly influenced by it (see Figure 31.8). Nutrient-rich watershed lead to nutrient-rich lakes and ponds. Excessive input of nutrients results in hypertrophy or pollution. Lakes and ponds also become settling basins for pesticides, herbicides, and toxic wastes contributed by the watershed. Nutrient-poor or oligotrophic watersheds usually support oligotropic lakes, but that condition can change with lake shore development. Students should be familiar with concepts of eutrophy, oligotrophy, and hypertrophy, especially as they relate to pollution.

### Source Materials

Among the titles in **Selected References,** see in particular Likens for a detailed analysis of the interaction of a lake ecosystem and its watershed, and the papers by Carpenter et. al. (1985), Carpenter and Kitchell (1984), and Carpenter (1980). R. G. Wetzel, 1975, LIMNOLOGY, Philadelphia: Saunders is an excellent treatment of open water life that does not ignore the littoral zone. The relationship between terrestrial and aquatic systems is reviewed in A. D. Hasler, ed., 1975, COUPLING OF LAND AND WATER SYSTEMS, New York: Springer-Verlag. Students may be directed to books on a more popular side , especially if they are unfamiliar with freshwater ecosystems. A good start is W. Amos, 1970, THE LIFE OF THE POND, New York: McGraw Hill. Two highly interesting British books are those by T. T. Macan (1963, 1970).

## Discussion Topics

1. Perhaps the most relevant discussion topics would be spin-offs from **Review and Study QUestion** 12. Some student probably will be familiar with highly eutrophic or polluted lakes in their own local areas that could serve for investigative reports. Conditions of urban lakes and ponds provide excellent material for the study of pollution and its effects on lake ecosystems.

2. In what ways could urban ponds and lakes be improved to make them more attractive places for recreation and wildlife? What economic benefits might accrue?

3. Does recreational use of lakes conflict with wildlife use? Consider the case of loons on northern lakes.

# CHAPTER 32

## FRESHWATER WETLANDS

### Commentary

If students are unaware of the controversy over what constitutes a wetland, a debate that reached the highest levels of government from the Office of the President to Congress they have not been reading newspapers, news and environmental magazines, or listening to news broadcast. The wetlands controversy is an excellent example of case in which scientific data gathered and used to define a wetland and then ignored by the very people who ordered the study. Students, like most people, do not consider a wetland to be one unless you can splash around in it. This attitude reflects a very narrow and erroneous concept of what constitutes a wetland. Wetlands include a variety of hydrological regimes, as indicated in Figure 32.1. It is important that students understand the range of wetlands and their importance.

Differentiation of various wetlands involves hydrology and hydroperiod, which affects their structure. Stress the role of hydroperiod on plant composition and wildlife use. Fresh meadows, for example, provide feeding grounds for migratory shorebirds; ephemeral ponds provide breeding places for amphibians; wooded swamps nesting sites for herons and wintering grounds for ducks. Hydroperiod involving periods of summer drawdown are essential for maintenance of wetland plants and the wetland itself. Students should appreciate all these nuances of wetland ecology. The role of hydroperiod peat differs in peatlands, including bogs and moors. The latter can develop when the hydroperiod of northern spruce forests is changed by logging. Refer back to discussion of boreal forest in Chapter 28.

Wetland functions are as variable as wetlands themselves, so you can consider only broad generalizations. Stress the seasonal role of biomass accumulation, nutrient cycling including late season deposits in roots and nutrient withdrawal early in spring, and decomposition rates of wetland detritus. These points are summarized in Figures 32.8, 32.9, and 32.10. Contrast nutrient cycling and decomposition between peatlands and other freshwater wetlands.

In part because wetlands are viewed by general public as wastelands and portrayed in fiction as places of evil and mystery, many people think they should be drained and the land converted to other uses. Because of this attitude, common to the Western World, wetlands have disappeared and are continuing to disappear at an alarming. Note the decline of wetlands in the United States as portrayed in Figure 32.12. The loss of wetlands contributed significantly to the disasterous decline in waterfowl, the sharp decline in amphibians, lowered water tables, increased flooding, among others. How to preserve our remaining wetlands is are a major environmental problem, so students should have a firm understanding of basic wetland ecology.

## Source Materials

Major source materials are given in **Selected References.** Students might be encouraged to review the wetland controversy by looking up articles in news and environmentally-oriented magazines. Relevant references include National Audubon Society (1990) and Tiner (1991). See also R. A. Leidy, P. L. Fiedler, and E. R. Micheli, 1992, Is wetter better? Bioscience 42(1):58-61, 65. That articles reviews the problems of using ecological data and theory in administrative policy relative to wetlands and other conservation issues.

## Discussion Topics

1. See **Review and Study Questions** 11, 12, 13, and 14, to which the following may be added:

2. If possible, visit a wetland in spring. Note the abundance and variety of wildlife. Particularly interesting would be an evening trip in spring when frogs are in chorus.

3. Discuss the paradox of one federal agency encouraging land drainage while another attempts to buy up wetlands to preserve them for wildlife.

4. Why does development to the edge of a wetland destroy the wetland?

5. Using Fish and Wildlife Service data, correlate the decline in waterfowl populations with the annual loss of wetlands.

# CHAPTER 33

## Flowing Water Ecosystems

### Commentary

Lotic systems are quite different from lentic systems and this chapter emphasizes those differences. This chapter on lotic systems considers physical characteristics and function. Note the differences in lotic systems from mountain brooks to river. There are major changes in forms of and adaptations to life along the continuum as well as heterotrophy and autotrophy. There is little discussion on adaptations to life in flowing water. Because water is constantly on the move, a problem in lotic systems is not only how to maintain position but also how to retain nutrients within any section of the system. The latter problem is solved in part by dependence on terrestrial inputs of nutrients, dependence largely on a detrital food chain, and retention of nutrients by spiraling. This important mechanism should be stressed in discussions of lotic systems, as well as the functional role of stream inhabitants--shredders, grazers or scrapers, collectors, and predators, and the associated food web as illustrated in Figure 33.6. If you do not get to this section during your course, you might introduce the concept in your discussions of nutrient cycling. Also be sure students are familiar with stream orders and with major detrital inputs: DOM, FPOM and CPOM.

Streams are endangered habitats. Pollution kills unique, oxygen-demanding organisms of fast streams. Siltation from housing developments, surface mines, road building, farmland erosion, and the like covers rubble bottom with silt and directly kills mayflies, stoneflies, fish, and other stream organisms. Dams turn flowing water ecosystems into lentic systems. Few streams are free-flowing any more. And streams that are left are channelized and straightened to hurry the water off. Productivity of channelized streams is only a fraction of natural streams. For an example see <u>California Fish and Game</u>, Vol. 62 (3): 179-186.

Chronic pollution is one problem faced by streams. Another, not discussed in the text, is the occasional or single event pollution that comes about when oil and toxic chemicals are released into a stream either by accident or deliberately. The pollutant moves downstream, killing all stream life as it passes by . The impacts of such events can have a long-lasting effect on the stream biota and structure.

The effects of pollution, channelization, and damming on fish species of flowing waters is enormous. Damming eliminates long stretches of riffle habitat. Siltation and pollution destroy both habitats and species. For example, at least 12 North American darters (<u>Percina</u> and <u>Etheostoma</u>) are threatened species and other face an uncertain future. Inhabitants of fast flowing water and riffles, darters are especially vulnerable to damming, pollution, and siltation.

## Source Materials

Among the **Selected References** see, in particular, Hynes (1970) and Ward (1979). A publication not listed in **Selected References** is J. R. Barnes and G. W. Minshall, eds., 1983, STREAM ECOLOGY: APPLICATION AND TESTING OF GENERAL ECOLOGICAL THEORY, New York: Plenum which reviews major ecosystem processes in flowing water except for secondary production. A good general introduction to flowing water is R. L. Usinger, LIFE IN RIVERS AND STREAMS, New York: McGraw Hill. W. Amos wrote two enjoyable books: 1970, THE INFINITE RIVER, New York: Random House, and 1981, WILDLIFE OF THE RIVERS, New York: Harry N. Abrams. A minor classic, one of my favorites, is J. E. Bardach, 1964, DOWNSTREAM: A NATURAL HISTORY OF A RIVER, New York: Harper and Row. The relationship between humans and flowing waters is well documented in R. H. Boyle, 1969, THE HUDSON RIVER, New York: W. N. Norton, available in paperback reprint editions.

## Discussion Topics

1. **Review and Study Questions 8, 9,** and 10 suggest topics for reports on human impacts on flowing water ecosystems.

2. What effects do dams have on the river continuum? Is the river continuum concept valid for regulated rivers? Why or why not? What effect do dams have on fish distribution?

2. What threatened or endangered species of fish or other flowing water inhabitants, such as freshwater mussels, are found in your region? What are the causes for the decline? What efforts are in place to protect the species?

# CHAPTER 34

## OCEANS

### Commentary

These last three chapters of the text are a highly concentrated survey of a very extensive topic, the marine ecosystems. None is given fair treatment, especially the open sea, but the subject is so large that many chapters would be needed to give a satisfactory survey. In a way this situation is unfortunate, because the oceans dominate Earth; they embrace an amazing diversity of lie, and are highly vulnerable to degradation by human activity. Not many of us give the oceans much though, even as we stand at the seashore or fly over them. THeir immensity is overwhelming and we can't quite comprehend the fact that oceans are living systems and not just huge basins of water.

This chapter first considers in a very brief manner the physical aspects of the marine environment: salinity, temperature and pressure, zonation and stratification, waves, currents, and tides, all important to the structure and functions of marine ecosystems. Students will discover that sea water is more than NaCl, that waves are generated far out at sea and the water that pounds on the rocks and sand is not water carried from some distant point.

This discussion is followed by a review of the structure of the open sea ecosystem in a very general way, because each ocean and region within an ocean supports its own distinctive structure and forms of life. Emphasis is on the biological: phytoplankton or primary producers, zooplankton consumers, nekton or larger free-swimming animals, the benthos or bottom-dwelling organisms, and the unique hydrothermal vents and their unique assemblage of life. The discussion on function centers on Marine productivity and energy flow, and food webs, summarized in Figures 34.4, 34.5, and 34.6. Of particular importance is the role of bacteria and protists in marine food webs. They give marine food webs a different look from those of terrestrial and fresh water food webs. Emphasize the importance of these organisms in marine ecosystems.

In spite of their immensity, marine ecosystems have been severely impacted by human activities. One of the major impacts is the pollution of oceanic waters; the other is the overexploitation of marine resources which can adversely impact marine food webs and the future of marine life. This topic is reviewed in some depth in Chapter 18. The effects of this degradation are experienced in one way or another, whether we find beaches closed by pollution or pay high prices for fish at the market and in resturants. Everyday experiences should make this chapter highly relevant to students. Those who read environmental news will be aware that the use of huge 30-mile long drift nets will be prohibited by international accord by 1994.

## Source Materials

Again major references are listed in **Selected References.** A major reference source for the physical aspects of the marine environment is O. Kinne, Ed., 1971, MARINE ECOLOGY, Vol. 1., ENVIRONMENTAL FACTORS (in three parts), New York: Wiley Interscience. Also consult D. H. Cushing and J. J. Walsh, eds., THE ECOLOGY OF THE SEAS, Philadelphia: Saunders, Marshall (1980) , and Steele (1974). More general books on the marine environment are M. G. Gross, 1982, OCEANOGRAPHY: A VIEW OF THE EARTH, Englewood Cliffs: Prentice-Hall for the physical side and J. W. Nybakken, 1988, MARINE BIOLOGY: AN ECOLOGICAL APPROACH, New York: Harper and Row. For more popular introduction to the seas, students should turn to such outstanding books as Rachael Carson, 1961, THE SEA AROUND US, New York: Oxford University Press; and A. Hardy, 1971, THE OPEN SEA: ITS NATURAL HISTORY, Boston: Houghton Mifflin.

## Discussion Topics

1. What has been the possible effect of sunken shipping and war fleets during the two World Wars on the marine environment? Might the source of the gobs of oil so noticeable on ocean surface come from the leaking fuel stores of sunken ships?

2. World fisheries are declining. As the best fish such as haddock are exploited to the point of extinction, fishing industry turns to species once regarded as trash fish such as turbot and monkfish. What happens when these species,too, are exploited to the point of economic extinction? What is the reason for the overexploitation of the sea? Why will the oceans not be able to feed the world as some optimists proclaim? What are the limits to productivity of seas? This question relates back to Chapter 18.

3. The demise of the krill-feeding fin whales, especially the blue whale, of the Antarctic seas, resulted in an reduction in competition for this food for penguins and other other Antarctic marine life. What might happen if the humans exploit krill for their own consumption? What effect would this exploitation have on the populations of penguins and seals? On the recovery of blue whale populations?

# CHAPTER 35

## INTERTIDAL ZONES AND CORAL REEFS

### Commentary

The seashore is perhaps the most fascinating of the marine environments. The first point to stress is the basic zonation of the shore. That is essential to the understanding the zonation of life on both rocky and sandy shores. Students should be familiar with the dominant organisms of each and their adaptations. Point out the gradation in shore life from north to south. Sand and mud shores are distinctly different environments from rocky shores. Although sandy shores appear to be barren, they exhibit a zonation similar to rocky shore. Sandy shores are dominated by detrital feeders buried in mud and sand. The pattern of life in the intertidal regions is heavily influenced not only by disturbance from wave action but also by the biotic interactions of grazing, predation, competition, and larval settlement. There is a lot more activity taking place among the inhabitants of rocky shores than is evident to the casual observer, and more life on the sandy beach than sunbathers realize. Stress the interaction of waves on community structure and productivity. Contrast conditions faced by inhabitants of rocky and sandy shores, and the role of bacteria in the food webs of sandy shores.

Discussion of coral reefs has been expanded from the previous edition. Stress the complexity of coral reef ecosystems and the mutualistic roles of the organisms inhabiting them, and the effects of disturbance, both natural and human on the structure of coral reefs.

Human impacts on the intertidal zone are intense. Because students are familiar with the intertidal zone, especially sandy beaches, they should respond to discussion of human activity on the intertidal zone. Polluted beaches are not uncommon, as evidenced by news reports of oil, solid wastes, and sewage washing up on shore and the sights of littered summertime beaches. Recreation and urban development has eliminated habitat critical to shore-nesting species, some on the verge of extinction, and reduced migratory stopover habitat for shorebirds. This chapter provides another opportunity of integrating ecological concepts to environmental problems: disturbance and competition; habitat loss and fragmentation on populations, and so on.

### Source Material

For a popular and accessible introduction to tidal zones, students should turn to such outstanding books as Rachael Carson's books, 1955 THE EDGE OF THE SEA, Boston: Houghton Mifflin , available in new reprint editions; N. J. Berrill, 1951 THE LIVING TIDE, New York: Dodd, Mead;  and C. M. Yonge, 1963, THE SEA SHORE, New York: Atheneum, also reprinted recently. From **Selected References** see Eltringham (1971) for sandy beaches and Stephenson

and Stephenson (1973) for rocky shores. A major descriptive reference is T. A>
Stephenson, 1972, LIFE BETWEEN THE TIDEMARKS ON ROCKY SHORES, SAN
FRANCISO: FREEMAN. J. Nybakken (1982, 1988:373-414) provides an
excellent introduction to coral reefs.

## Discussion Questions

1. The eastern piping plover, nesting only on sandy beaches and open sandy
areas on the dunes, has been reduced to a few small isolated populations and is
in great danger of extinction. How has human use of the sandy beaches helped
to drive this bird to its current condition?

2. In contrast the human use of the seashore, providing an abundance of food in
the form of garbage, has stimulated a rapid population of growth of highly
adaptable herring gull and black-backed gull. What effect has the population
explosion of these two species had on other nesting coastal bird, especially the
terns?

3. How did the offshore dumping of sewage in the Hawaiian Islands affect some
coral reefs?

4. What effect does the exploitation of coral for decorative use and jewelry
have on the coral ecosystems as well as the coral species? The United States is
the largest consumer of raw coral.

# CHAPTER 36

## ESTUARIES, SALT MARSHES, AND MANGROVE FORESTS

### Commentary

This last chapter considers estuaries, the place where fresh water meets the sea. This mixing of fresh water with salt creates a unique environment where the physical structure is strongly influenced by differences in salinity and temperatures. To appreciate the estuary, students must have some understanding of the physical aspects of it, particularly salinity gradients and the interaction of outflowing fresh water and inflowing salt water. This interaction creates a nutrient trap, responsible for the richness and high productivity of the estuarine ecosystem. The nutrient trap, in a way, is the estuarine counterpart of spiraling in stream ecosystems. Both are associated with flowing water; both are mechanisms for keeping nutrients in place in spite of it. Compare Figures 36.3, a generalized model of the nutrient trap with Figure 36.4. a more detailed model.

Associated with estuaries are tidal or salt marshes. Figure 36.6 illustrates the zonation of salt marshes as influenced by salinity and tides. A great deal of variation, however, exists among salt marshes not only in structure, but also in function. Note the differences between salt marshes of northern and southern North America. Although the discussion is centered on salt marshes of eastern North America, because they are the best developed and studied from a functional point of view, salt marshes of other parts of the world, especially western Europe, deserve some attention. Emphasize the role of salt marshes in the estuarine system and their importance to coastal wildlife and to estuarine fishes.

Mangrove forests replace salt marshes in tropical regions. In North America they make their first appearance in southern Florida. Elsewhere they form the major vegetation of tropical coastlines. Mangrove forests are extremely important both ecologically and economically. More so than salt marshes, mangroves export a considerable portion of their production to surrounding water, important to coastal finfish and shellfish.

Because they are coastal, where human settlement are concentrated, the estuary and its associated salt marshes and mangrove forests are vulnerable to human disturbance and destruction. Estuaries receive not only the inputs from coastal development but also all the inland pollutants carried by inflowing rivers. Salt marshes are prime targets for development from retirement villas to marinas. Mangroves are eliminated for agriculture and for wood products. Severely affected are the dependent estuarine fisheries and wildlife. The intense development of coastal regions of North America provide the basis for in-depth discussion of their effects on the future of coastal ecosystems.

## Source Material

For the estuarine and coastal systems, including tidal mashes, see Jefferies and Davy, (1979), McLusky (1981), Perkins (1974), Wiley (1976), Ketchum (1983), a major reference, and one not listed, K. Mann, 1982, ECOLOGY OF COASTAL WATERS: A SYSTEMS APPROACH, Berkeley: University of California Press. The biology of estuarine waters is neglected in the text. For information consult J. Green, 1968, THE BIOLOGY OF ESTUARINE ANIMALS, Seattle: University of Washington Press and R.Remane and C. Schlieper, 1971, BIOLOGY OF BRACKISH WATERS, New York: Wiley. Providing a synthesis of information on structure and function of estuaries, including the ecology of key organisms, is J. W. Day et al., 1989, ESTUARINE ECOLOGY, New York: Wiley. Major references for salt marshes include Chapman (1977), Pomeroy and Wiegert (1981), and Long and Mason (1983), the latter an excellent introduction with an European approach. See Odum et al. (1982) for a detailed study of function of mangrove ecosystems, including food webs, in southern Florida. An excellent reference on mangroves is Chapman (1976 and 1977).

For a popular introduction to salt marshes and estuaries,, students should turn to J. Teal and M. Teal, THE LIFE AND DEATH OF A SALT MARSH, Boston: Little Brown; and W. W. Warner, 1976, BEAUTIFUL SWIMMERS: WATERMEN, CRABS, AND CHESAPEAKE BAY, Boston: Little Brown.

## Discussion Questions

1. **Review and Study Questions 13, 14, 15,** and **16** provide a basis for reports on problems of the estuarine systems. Two other questions can be added to those:

2. What is the fate of our coastal marshes? What major cities are built on filled coastal marshes? Why did such "reclamation" take place so early in or history? To what extent are states working toward the protection of the coastal wetlands or are they succumbing to the "development syndrome?"

3. For an example of a major estuary in trouble, consider the Chesapeake Bay or San Francisco Bay. What are the problems and what steps are being taken to save the Bays?

# TEST BANK

## Chapter 1

## What is Ecology?

### SENTENCE COMPLETION

1.1   The study of freshwater environments is called **limnology**.

1.2   Malthus proposed the idea that populations grow in a
      **geometric** fashion and **double** at regular intervals.

1.3   The subdivision of ecology concerned with how populations
      grow is **population** ecology.

1.4   **Physiological** ecology examines the responses of organisms
      to abiotic environments, such as temperature, moisture, and
      light.

1.5   **Sociobiology** holds that behavior is genetically controlled.

1.6   The specialized field of **chemical** ecology examines the
      nature of chemical substances in the natural world.

1.7   A(n) **hypothesis** is a statement about an observation that
      can be tested experimentally.

1.8   Formulas for population growth and predation are examples
      of **analytical** models.

1.9   **Simulation** models are useful in examining complex
      environmental interactions, such as energy flow and
      nutrient cycling.

### MULTIPLE CHOICE

1.10  "If you can discover how each part of a system functions
      you will know how the whole operates."  This statement best
      describes:
         a) a holistic approach to ecology;
         b) an ecosystem approach to ecology;
         **c) a reductionist's approach to the study of ecology;**
         d) an approach to ecology that thinks ecological
            systems are too complex to study in isolated bits
            and pieces;
         e) an approach to ecology that falls more preciously
            into the realm of evolutionary ecology.

105

1.11 Environmental studies is an academic discipline that draws heavily on ideas developed in the areas of:
a) ecology;
b) geology;
c) economics;
d) economics;
**e) it draws information from all of the above.**

1.12 The terms "producers" and "consumers" are most often associated with the investigations of:
a) Niko Tinbergen;
b) Gregor Mendel;
**c) A. Thienemann;**
d) F.C. Clements;
e) R.A. Lindeman.

1.13 Which of the following scientists called the public's attention to the environmental problems created by pesticide use?
**a) Rachel Carson;**
b) R. F. Chandler;
c) Aldo Leopold;
d) Paul Erlich;
e) Garrett Hardin.

1.14 Which of the following statements is incorrect?
a) experimentation involves manipulation to test a hypothesis;
b) evolutionary biology allows scientists to study the effects of environmental variables on natural selection;
c) in order to understand the dynamics of the natural world we must depend on research from both population ecology and ecosystem ecology;
**d) most ecological experiments are conducted in the field.**

TRUE/FALSE

1.15 **T** F Ecology is a science that studies the relationship between organisms and their environment.

1.16 **T** F Environmental science is an interdisciplinary subject that combines ideas from many sciences.

1.17 T **F** Physiological ecology examines how genetics controls behavior.

1.18 **T** F   A holistic ecologist would agree with the statement that ecosystems are too complex to study in small fragments, but should be studied as large functional groups.

1.19 **T** F   As the number of prey increases in the home range of some predatory birds, the number of eggs these birds lay increases. This observation was probably made by a descriptive ecologist.

1.20 T **F**   Ecologists always examine the subject matter in natural settings; they never depend on laboratory experimentation.


**MATCHING**

1.21 __**H**__   Examined the role of imprinting and instinct on the social life of animals.

1.22 __**B**__   Plant succession.

1.23 __**C**__   Ideas of organic nutrient cycling and trophic feeding levels.

1.24 __**A**__   Similar climates support similar vegetation.

1.25 __**D**__   Integrated food cycle dynamics with the principles of community succession.

1.26 __**F**__   Examined the responses of plants to limited supplies of nutrients.

1.27 __**E**__   Population grow in a geometric fashion and double and regular intervals.

1.28 __**G**__   Described the relationship between plants and animals and developed the concept of biotic communities.


A.  Carl Willdenow        B.  F.C. Clenents
C.  A Thienemann          D.  R.A. Lindeman
E.  Thomas Malthus        F.  Justus Von Liebig
G.  Victor Shelford       H.  Konrad Lorenz

## DISCUSSION

1.29    The term "ecology" is most often confused with the term "environmental science". How are these two sciences similar and how are they different?

1.30    Ecologists cannot give definitive answers. Do you agree or disagree with this statment? Explain your answer.

1.31    Modern ecology has developed along two major lines, ecosystem ecology and population ecology. How does each of these disciplines differ in their approach to examining the environment?

## Chapter 2

## Natural Selection and Evolution

<u>**SENTENCE COMPLETION**</u>

2.1 Geologists estimate that the earth is **4600 million** years old.

2.2 **Variation** is the raw material of natural selection.

2.3 The **fitness** of an individual is measured by its ability to produce healthy offspring.

2.4 A **deme** is a local population or interbreeding group within a larger population.

2.5 The sum of the hereditary material an organism carries of its **genotype.**

2.6 A bird has red feathers.  This observable expression of a genetic code is its **phenotype.**

2.7 **Phenotypic plasticity** refers to the ability of a genotype to cause many phenotypic expressions.

2.8 The position of a gene on a chromosome is its **locus.**

2.9 **Mutations** are changes in a genetic code that are inheritable.

2.10 When a small segment of a population splits off from the main population phenotypes in the small population quickly change.  This phenomena is called the **founder effect.**

2.11 The elimination of a genetically inherited factor simply by chance is **genetic drift.**

2.12 The fact that most flowering plants are pollinated by insects is an example of **co-evolution.**

2.13 The sum of all the genes in a population is the **gene pool.**

2.14 Darwin could not explain completely the mechanisms of evolution because he was unfamiliar with the mechanisms of **inheritance.**

2.15 The unit of evolution is the **population.**

2.16 In natural selection, **phenotypes** are selected for or against, and **populations** evolve.

2.17 Hardy and Weinberg described the frequency of **alleles** for an entire population.

2.18 If the frequency of one allele is p, then the frequency of the alternative allele in the population is **1-p**.

2.19 When and where a mutation occurs depends on **chance**.

2.20 Close inbreeding within a population can result in **inbreeding depression**.

## MULTIPLE CHOICE

2.21 What is the currently estimated age of the earth?
   **a) 4.5 billion years;**
   b) 3 million years;
   c) 2000 years;
   d) no estimate has been made.

2.22 A change in the overall genetic makeup of a population is:
   a) inheritance of acquired characteristics;
   b) natural selection;
   c) founder effect;
   **d) evolution.**

2.23 Inherited mutations are produced:
   a) as a response to environmental changes;
   b) as a response to selective pressures;
   c) to enable a population to survive environmental change;
   **d) solely by chance.**

2.24 Natural selection depends on the interaction between an organism and its:
   **a) environment;**
   b) genes;
   c) phenotype;
   d) offspring.

2.25 The genetic makeup of an individual is called:
   a) phenotype;
   **b) genotype;**
   c) gene pool;
   d) mutation.

2.26 The fundamental unit of evolution is the:
   a) gene;
   **b) population;**
   c) individual;
   d) species.

2.27 All of the following meet the Hardy-Weinberg requirements
     for equilibrium in a population except:
         **a) no random mating can occur;**
         b) no mutations may occur;
         c) no migration may occur;
         d) no natural selection may occur.

2.28 Bottlenecks:
         a) occur when a population increases in size;
         b) result in an increase in the diversity of available
            alleles in the gene pool;
         c) occur because populations naturally fluctuate as
            environmental conditions change from year to year;
         **d) may prevent a species from reversing its path to
            extinction.**

2.29 In a hypothetical population, 80% of the population carries
     a dominant allele, B, and 20% of the population carries a
     recessive allele, b.  What percentage of the population
     would you expect to be heterozygous?
         a) 4%;
         b) 16%;
         c) 32%;
         **d) 64%.**

2.30 Adaptation:
         a) is any form of behavior that is the result of
            natural selection;
         b) is a change in the physical, physiological, or
            behavioral traits that result from some
            environmental pressure;
         c) is any physical or physiological feature used to
            explain the ability of an organism to live in a
            particular environment.
         **d) all of the above answers are correct;**
         e) none of the above answers are correct.

2.31 Death of extremely overweight and extremely underweight
     babies is an example of:
         a) directional selection;
         **b) stabilizing selection;**
         c) disruptive selection;
         d) survival of the fittest.

2.32 Charles Darwin probably would
have agreed with which of the following statements:
   a) giraffes have had long necks since the time of their
   creation;
   b) giraffes developed long necks by reaching to the
   tops of trees to eat leaves;
   c) giraffes have undergone several dramatic changes in
   the body structure since their creation;
   **d) long-necked giraffes could find more food at the top
   of trees and survived to pass the trait on to their
   offspring.**

2.33 The nitrogen base cytosine always pairs with:
   **a) guanine;**
   b) thymine;
   c) adenine;
   d) cytosine.

2.34 Which of the following statements is correct about mutations
of single genes?
   a) they are not important because they reduce variation
   in the gene pool;
   b) they direct evolutionary change in populations;
   c) they reduce the size of the gene pool;
   **d) they restore and maintain variation in the gene
   pool.**

2.34 Gametes are formed by the process of:
   **a) meiosis;**
   b) mitosis;
   c) recombination;
   d) mutation.

2.35 Variations within populations arise from:
   a) gene mutations which generate new alleles;
   b) fertilization between genetically varied gametes;
   c) chromosomal aberrations;
   d) a and b, but not c.
   **e) a, b, and c are correct.**

2.36 Inbreeding:
   a) is the result of mating between widely separated
   demes;
   b) decreases homozygosity and increases heterozygosity;
   **c) is often detrimental because rare, recessive genes
   become expressed;**
   d) favors the heterozygote condition.

2.37 Which of the following statements is true?
  a) Inbreeding involves random mating;
  **b) In polygamous populations the ratio of breeding males and females are unequal;**
  c) In a monogamous population the offspring are more related than in a polygamous population;
  d) If immigrants enter a breeding population, they will speed up genetic drift and thus reduce genetic diversity.

2.38 Which of the following conditions has resulted in a reduction of genetic variability in populations?
  a) habitat fragmentation;
  b) poaching;
  c) increasing human numbers;
  **d) all of the above;**
  e) none of the above.

2.39 Mimicry in insects may result in:
  a) directional selection;
  **b) disruptive selection;**
  c) stabilizing selection;
  d) balanced polymorphism.

**TRUE/FALSE**

2.40 T **F** Variations in characteristics that result from disease or injury are inheritable.

2.41 **T** F Polyploidy, the duplication of entire sets of chromosomes may arise from an irregularity in meiosis.

2.42 T **F** Allopolyploidy is more common in animals than plants because animals depend on cross fertilization.

2.43 **T** F The diploid state Aa is a heterozygote.

2.44 T **F** The assumptions operating in the Hardy-Weinberg Law apply to all natural populations.

2.45 **T** F Evolution is a change in a gene frequency through time.

2.46 **T** F Most changes in biological systems are due to factors in the environment such as predation, food availability and competition.

2.47 **T** F Bottlenecks occur when a large population is drastically reduced in size.

2.48  T  **F**    The fact that many insects are resistant to pesticides is used as an example of disruptive selection.

2.49  **T**  F    Early in Earth's geologic history continents changed position and orientation.

2.50  T  **F**    Prokaryotes have nuclei and chromosomes.

2.51  **T**  F    The fact that siblings can recognize themselves over time serves to reduce close inbreeding.

2.52  T  **F**    In the formation of the rungs of the DNA molecule, adenine pairs with guanine.

2.53  T  **F**    Phenotypic plasticity refers to variations in phenotypic expressions as a response to varying environmental conditions.

2.54  T  **F**    Natural selection acts on an organism's dominant alleles only.

2.55  **T**  F    Genetically based advantageous phenotypes tend to increase in frequency.

2.56  **T**  F    Individuals in populations with more adaptive traits have a greater chance of leaving more offspring.

2.57  **T**  F    Chromosomes are composed of DNA.

2.58  **T**  F    Founder effect is a special case of genetic drift and is more common among populations on islands.

## MATCHING

If the frequency of A is 0.9 and that of a is 0.1, then the genotype frequencies are: AA____, Aa____, aa____.

2.59  __C__    AA

2.60  __A__    Aa

2.61  __D__    aa

         A.  18%
         B.  12%
         C.  81%
         D.  1%
         E.  10%

2.62 __C__ Selection that favors individuals at the extremes of the population instead of at the mean.

2.63 __B__ Selection that occurs because of a response to a rapidly changing environment.

2.64 __A__ Selection that favors phenotypes near the population mean rather than at its two extremes.

    A. Stabilizing slection;
    B. Directional selection;
    C. Disruptive selection.

2.65 __B__ This geologic era is characterized by the origin of the first living cells.

2.66 __D__ Land plants first appear during this geologic era.

2.67 __A__ This era is characterized by the dominance of dinosaurs and flowering plants.

2.68 __C__ During this era placental mammals began their evolution in North America.

2.69 __E__ Geologic period that ushers in the beginning of the Ice Age.

2.70 __F__ During this geologic period temperature began to cool and reptiles began to disappear.

    A.  Mesozoic        B.  Precambrian
    C.  Eocene          D.  Paleozoic
    E.  Pleistocene    F.  Cretaceous

## DISCUSSION

2.71     What happens to the genetic variability of small, isolated populations of laboratory organisms after many generations without the introduction of new individuals to the population?

2.72     How are new alleles created? Is the creation of new alleles an important source of genetic change?

2.73     List five sources of phenotypic variation.

2.74     Why are conservationists concerned when the genetic variation within a population of rare or endangered organisms begins to decrease?

# Chapter 3

## Natural Selection and Speciation

### SENTENCE COMPLETION

3.1 Populations separated by a geographic barrier are known as **allopatric** populations.

3.2 A **species** consists of many local populations upon which natural selections acts.

3.3 **Sympatric** speciation occurs most often in plants.

3.4 Two species of pines occur in the same habitat and can hybridize. One pine releases pollen in February and the other in April. They are examples of **sibling** species.

3.5 A gradual change in the characteristics of a population over a geographic area generally as a result of changes in environmental conditions is referred to as a **cline.**

3.6 **Linnaeus** developed a standardized method of naming living organisms.

3.7 The **kingdom** is the most general group used in Linnaeus' classification hierarchy.

3.8 In binomial classification <u>Felis</u> <u>concolor</u>, "felis" signifies the **genus.**

3.9 An **ecotype** is a genetic strain of a population that is adapted to a set of local environmental conditions.

3.10 **Polytypic** ecotypes have evolved independently from different local populations.

3.11 A **subspecies** is an aggregate of local populations differing taxonomically from other populations of the same species.

3.12 The formation of many species from a single common ancestor is called **adaptive radiation.**

3.13 The occurrence of distinct forms of a species in the same habitat at the same time is **polymorphism.**

3.14 Differences in courtship and mating behavior in animals act as **ethological** barriers keeping species distinct.

3.15 Birds and many fish depend on **visual** signals to local the correct mate.

3.16 If genes from hybrids are retained and incorporated into the gene pool **introgressive hybridization** has taken place.

3.17 Two duck populations are genetically different but these genetic differences are not sufficient to prevent interbreeding. The two populations are **semispecies**.

3.18 Two populations that develop similar characteristics in response to similar environments show **parallel** evolution.

3.19 "Most species arise from a slow transformation from an ancestral species". Such an explanation for evolutionary change is called **gradualism**.

3.20 **Punctuationalism** suggests that species arise instantaneously during times of geological instability.

## MULTIPLE CHOICE

3.21 A premating isolating mechanism prevents successful:
  a) reproduction of hybrids;
  b) zygote development;
  **c) fertilization;**
  d) gamete production.

3.22 Two species of frogs are found in the same geographic area, but one lives in a pond, the other in a stream. They do not encounter one another and so they do not interbreed. This situation can serve as an example of a premating isolation caused by:
  **a) habitat isolation;**
  b) behavioral isolation;
  c) temporal isolation;
  d) mechanical isolation.

3.23 Whenever species evolve as contiguous populations in a continuous cline, without physical barriers to gene flow _____ speciation is taking place.
  **a) parapatric;**
  b) allopolyploid;
  c) autopolyploid;
  d) sympatric.

3.24 The correct sequence from a hierarchical classification is:
  a) kingdom, phylum, genus, order, family;
  b) kingdom, class, phylum, order, family, genus, species;
  **c) kingdom, phylum, class, order, family, genus, species;**
  d) kingdom, phylum, order, class, family, genus, species.

3.35 A postmating isolating mechanism prevents successful:
  a) courtship;
  b) copulation;
  c) fertilization;
  **d) development, survival, or reproduction of a hybrid.**

3.36 In allopatric speciation, the initial barrier to gene flow is:
      a) behavioral;
      b) mechanical;
      **c) geographic;**
      d) genetic.

3.37 Populations of Kaibab squirrels on the North rim of the Grand Canyon do not mate with populations on the South rim. The two populations are isolated by a:
      a) sympatric barrier;
      b) autopatric barrier;
      c) genetic barrier;
      **d) geographic barrier.**

3.38 The most important isolating mechanisms in animals are:
      **a) ethological barriers;**
      b) mechanical barriers;
      c) ecological barriers;
      d) geographic barriers.

3.39 Members of a species are able to:
      **a) interbreed;**
      b) compete;
      c) adapt;
      d) co-exist.

3.40 The formation of two species from one ancestral species is called:
      **a) divergent evolution;**
      b) co-evolution;
      c) parallel evolution;
      d) polyploidy.

3.41 Two animals with a common evolutionary heritage develop similar body parts in response to similar environmental pressures. Such an event is called:
      a) convergent evolution;
      b) adaptive radiation;
      c) divergent evolution;
      **d) parallel evolution.**

3.42 Some species reduce their differences and become similar. This phenomena is called:
      **a) convergent evolution;**
      b) parallel evolution;
      c) divergent evolution;
      d) adaptive radiation.

3.43 Eastern and Western meadowlarks look very similar and live
     in the same habitat.  They recognize mates of their own
     species by distinctive songs.  This is an example of:
     a) ecological isolation;
     b) gametic isolation;
     c) mechanical isolation;
     **d) behavioral isolation.**

3.44 Punctuationalism:
     a) asserts that most change involves speciation in
        large populations;
     b) is marked by instantaneous speciation followed by
        very short periods of species stability;
     c) was a central idea in Darwinian evolution;
     **d) suggests that new species arise by total
        transformation of a linage rather than by slow
        adaptation.**

3.45 Species arise by the interaction of:
     a) heritable variation;
     b) natural selection;
     c) barriers to gene flow;
     d) a and b, but not c;
     **e) a, b, and c.**

3.46 An outcome of speciation is:
     **a) adaptive radiation;**
     b) parallel evolution;
     c) convergent evolution;
     d) divergent evolution.

3.47 Instantaneous speciation, especially in plants, is the
     result of:
     a) adaptive radiation;
     b) parapatric speciation;
     **c) polyploidy;**
     d) introgressive hybridization.

3.48 Adaptive radiation has occurred often on islands, because
     island populations are:
     a) without predators;
     **b) are small in number and isolated;**
     c) asexual;
     d) hybrids.

3.49 Many different species of fishes release their sperm and
     eggs in the same water, but the sperm of one species will
     not penetrate the eggs of another species.  This example
     illustrates:
     a) ecological isolation;
     **b) temporal isolation;**
     c) behavior isolation;
     d) geographic isolation.

3.40 An example of adaptive radiation is:
         a) pesticide resistant insects;
         b) industrial melanism;
         **c) Darwin's finches;**
         d) bird songs.


## TRUE/FALSE

3.41  **T**  F     Two allopatric populations can interbreed once the isolating barrier is removed.

3.42  **T**  F     The greater the distance between populations the greater the variation between the two population.

3.43  **T**  F     Populations at the two extremes along a geographic gradient may behave as separate species.

3.44  T  **F**     Clinal gradation is characteristic of animal species but not plant species.

3.45  **T**  F     Variations in the pigmentation of the moth <u>Biston betularia</u> is an example of polymorphism.

3.46  T  **F**     In the industrial area of England the light melanistic form of the White Peppered Moth is more common than the dark phase.

3.47  T  **F**     Reproductive isolation in sympatric speciation requires a geographic barrier.

3.48  **T**  F     Hybrids are often sterile and are quickly eliminated from the breeding population.

3.49  T  **F**     In founder populations selection favors heterozygosity.

3.50  **T**  F     The prerequisite for sympatric speciation is the formation of a stable polymorphism through disruptive selection.

3.51  T  **F**     Parapatric speciation is most commonly found among populations that are reproducing slowly.

3.52  **T**  F     Sympatric speciation is most likely to occur among insect parasites of plants and animals.

3.53  **T**  F     Punctuationalism suggests that species are rapidly established during times of environmental instability.

3.54  T  **F**     Evolution and speciation always take place rapidly.

3.55  **T**  F    Evolution among some populations is always taking place somewhere on Earth.

3.56  **T**  F    Parapatric speciation is more common among plants than animals.

3.57  T  **F**    A biological species is a group of interbreeding organisms living together in the same environment.

3.58  **T**  F    The most common method by which abrupt speciation takes place is by polyploidy.

3.59  T  **F**    Adaptations that allows an organism to become established in a new environment become "fixed" and are rarely altered by selection.

3.60  T  **F**    The development of superifically similar structures by unrelated forms is coevolution.

## MATCHING

3.61  __B__    Two species of houseflies do not mate with each other because the males of each species have appendages that can hold, for copulation, only females of their species.

3.62  __C__    Male fireflies attract mates by blinking; each species has a characteristic pattern of blinks that are recognized only by females of their species.

3.63  __D__    Two species of oak trees are capable of hybridization.  They do not interbreed because each releases its pollen at a different time.

3.64  __A__    Very similar species of crabs release their eggs and sperm into the same tidepool, but sperm of one species cannot penetrate the eggs of another species.

> A.  gametic isolation
> B.  mechanical isolation
> C.  behavioral isolation
> D.  temporal isolation

3.65  __C__    The dark form of <u>Biston betularia</u> is limited in range to industrial areas of England.

3.66  __A__    A new flower shape appears in a population and soon it completely replaces the original form so that the original form can only be maintained by recurrent mutation.

3.67 __B__    Swallowtail butterflies exhibit two pupa color
              phases depending on the color of the background
              environment they select.

3.68 __C__    Environmental changes convert a mutant allele into
              a beneficial one.

3.69 __B__    Environmentally controlled forms of polymorphism
              favoring two or more forms.

3.70 __B__    In land snails there are several forms of shell
              color. Each form is well-camouflaged from
              predators active within a particular
              microenvironment.

        A.  transient polymorphism
        B.  stable polymorphism
        C.  genetic polymorphism

## DISCUSSION

3.71 What is the difference between a morphological species and a
     biological species?

3.72 What evidence can you present to support the ideas of
     gradualism and punctuationalism? Which explanation do you
     believe is the most logical? Why?

3.73 Why is sympatric speciation common in plants but rare in
     animals?

3.74 Describe the most common sequence of events that will lead
     to animal speciation. How does this sequence differ for
     plants?

3.75 Why would you expect that some animals, such as ducks, have
     elaborate courtship behaviors, whereas other animals do not?

# Chapter 4

## Climate

**SENTENCE COMPLETION**

4.1  **Homeostasis** is the tendency of a system toward maintenance of a condition nearly constant in the face of a changing environment.

4.2  **Negative feedback** is a mechanism by which the change detected  in some condition stimulates compensating physiological activities that bring the condition back to within its normal range.

4.3  **The Law of the Minimum** says that the well-being of an organism is limited by that resource that is in lowest supply relative to what is needed.

4.4  **The Law of Tolerance** says that for each physical factor in an environment, a minimum and maximum limit exist, beyond which no member of a species can survive.

4.5  **Weather** refers to the temperature, humidity, winds, and other conditions at a given place and time.

4.6  The summation of weather conditions over a long period is **climate.**

4.7  1.94 cal/cm2/min is known as the **solar constant.**

4.8  Some radiation from the sun is absorbed by Earth and is emitted back to the atmosphere as **long-wave radiation.**

4.9  Gases in the atmosphere, especially **carbon dioxide** and water vapor trap much of the energy from the sun.

4.10 The absorption and reradiation of infra red radiation by gases in the atmosphere is called the **green house effect.**

4.11 The Earth is tilted at **23** degrees to the plane of its  orbit about the sun.

4.12 An **inversion** is an atmospheric condition in which the temperature of the air increase with height.

4.13 When cool, moist air from the ocean spreads over warm land a **marine inversion** is formed.

4.14 The **Coriolis effect** acts on north-south wind currents to give them an east-west component.

4.15 The **intertropical convergence** causes a rainy season in the tropics and a dry season in savannas.

4.16 The moisture content of the air is expressed as **relative humidity.**

4.17 The **dew point** is the temperature at which atmospheric water condenses.

4.18 The **microclimate,** not the local climate defines the conditions under which most organisms live.

4.19 You would expect that plants on the **north-facing** slope of a hill would prefer moist microclimates.

4.20 When gases in the atmosphere **expand** they cool, but when they **compress** they heat.

4.21 **General winds** move with the leading edge of an air mass, whereas convective winds are caused by changes in the local temperatures.

4.22 The Pacific and Atlantic Ocean exhibit great circular water motions, or **gyres**

4.23 If the vegetation in a locality is removed the temperature near the soil surface would become **warmer.**

4.24 Human activities, such as power production and industrial combustion **raise** the temperature of Earth.

4.25 Dust and other particulate material can become **condensation nuclei** upon which water vapor will precipitate.

## MULTIPLE CHOICE

4.26 Which statement is true?
  a) **climate is the general pattern of weather over several years;**
  b) climate is the weather condition at a given time and place;
  c) climate will change in a given area from day to day;
  d) weather is a more important limiting factor to plants than climate.

4.27 Of the incoming solar radiation from the sun approximately _____ percent is reflected by the atmosphere and by Earth.
  a) 60;
  b) 10;
  c) **30;**
  d) 50.

4.28 Of the incoming solar radiation approximately
_____percent is absorbed by the atmosphere and by Earth.
      **a) 70;**
      b) 60;
      c) 30;
      d) 50.

4.29 The atmospheric gas_____ is responsible for trapping
incoming energy from the sun.
      a) oxygen;
      b) nitrogen;
      c) hydrogen;
      **d) carbon dioxide.**

4.30 Which statement is true?
      a) during a green house effect atmospheric temperature
         decreases;
      b) during a green house effect  atmospheric temperature
         near the ground is warmer than above the ground;
      c) during a green house effect winds are generated and
         the atmosphere is cleansed;
      **d) during a greenhouse effect the atmosphere absorbs**
         **outgoing long-wave radiation and reflects it back**
         **toward Earth.**

4.31 The sun's energy does not reach Earth uniformly because:
      a) the shape of Earth influences the amount of energy
         reaching the surface;
      b) the Earth tilts on its axis;
      c) the elliptic orbit of Earth about the sun influences
         the concentration of solar energy;
      d) a and b, but not c;
      **e) a, b, and c.**

4.32 The average air temperature in the tropics remains constant
because:
      a) heat moves from high latitudes to the tropics;
      b) there is an erratic fluxuation of solar radiation at
         low latitudes;
      c) low latitudes experience a greater variation in
         incoming solar radiation;
      **d) heat generated in the tropics moves to high**
         **latitudes, because there is a steady solar radiation**
         **at low latitudes;**

4.33 When cool, moist air masses move from the ocean over land a
_____ is created.
      a) subsidence inversion;
      **b) marine inversion;**
      c) radiation inversion;
      d) surface inversion.

4.34 Which statement is correct?
    a) cool, light air sinks over arctic regions, then flows toward the equator;
    b) cool, heavy air rises over the tropics and flows to the poles;
    c) warm, light air rises over the arctic regions and flows toward the equator;
    **d) cool, heavy air sinks over the arctic and flows toward the equator.**

4.35 Because of the Coriolis effect:
    **a) Northern Hemisphere airflow is directed to the right and in the Southern Hemisphere to the left;**
    b) Northern Hemisphere airflow is directed to the left and in the Southern Hemisphere to the right;
    c) Northern Hemisphere airflow is directed to the right and in the Southern Hemisphere to the right;
    d) Northern Hemisphere airflow is directed to the left and in the Southern Hemisphere to the left.

4.36 Each hemisphere has_____ cells of airflow.
    a) two;
    b) five;
    **c) three;**
    d) six.

4.37 Gyres are:
    **a) circular ocean currents;**
    b) tropical environments;
    c) localized wind patterns;
    d) doldrums.

4.38 _____are the wind patterns found near the equator.
    a) tradewinds;
    b) westerlies;
    c) easterlies;
    **d) doldrums.**

4.39 The important gas in the green house effect is:
    **a) carbon dioxide;**
    b) nitrogen;
    c) oxygen;
    d) hydrogen.

4.40 High pressure cells are associated with:
    a) clouds and rain;
    b) clouds and dry air;
    **c) clear sky and dry air;**
    d) clear sky and high winds.

4.41 What type of environment would you expect at 60 degrees
     north latitude and 60 degrees south latitude?
        a) deserts;
        b) forests;
        c) grasslands;
        **d) arctic.**

4.42 Which of the following statements is false?
        a) microclimate rather than local climate defines the
           conditions in which most organisms live;
        b) soil is an active surface that can absorb solar
           radiation;
        c) heat absorbed by the soil during the day is
           reradiated by the soil at night;
        **d) if the air above the ground is dry very little heat
           leaves the soil and the surface of the ground warms
           up.**

4.43 Vegetation influences microclimate by altering:
        a) wind movement;
        b) soil temperature;
        c) evaporation;
        d) a and b, but not c;
        **e) a, b, and c.**

4.44 Which of the following statements is true?
        **a) vegetation influences the height of the active
           surface;**
        b) without vegetation temperature decreases
           near the soil;
        c) in dense vegetation temperatures are higher at the
           surface of the ground and lowest at the plant
           crowns;
        d) low temperature and high vapor pressure will force
           moisture out of the soil and plants.

4.45 Human-caused changes in the atmosphere may result in:
        a) global warming:
        b) rise in the levels of the ocean;
        c) lengthening of the growing season at high
           latitudes;
        d) an increase in atmospheric temperature in cities;
        **e) all of these events may occur.**

**TRUE/FALSE**

4.46  T  **F**    Air in mountain valleys is cooler during the day
                  than at night.

4.47  **T**  F    You would expect timberline plants to be adapted
                  to dry environments.

4.48  **T**  F      If air is warmed while its moisture content remains constant, relative humidity drops.

4.49  T  **F**      Relative humidity decreases with increase in altitude.

4.50  T  **F**      For most organisms average annual precipitation is more important than seasonal distribution of rainfall.

4.51  **T**  F      Currents in the Northern Hemisphere's gyre move clockwise, and counter-clockwise in the Southern Hemisphere.

4.52  T  **F**      Currents moving toward the equator along the west sides of continents will result in warm, dry climates.

4.53  **T**  F      Vegetation alters wind movement, moisture and soil temperature and as a result microclimates are created.

4.55  T  **F**      North-facing slopes receive more sunlight than south-facing slopes.

4.56  **T**  F      High temperatures and low vapor pressure will result in dry environments.

4.57  T  **F**      You would expect to find woodlands on the north-facing slope of a mountain.

4.58  **T**  F      The average temperature in a city is greater than in the surrounding open countryside.

4.59  **T**  F      Climate is the summation of all the weather conditions over a long period.

4.60  T  **F**      Most of the energy from the sun that strikes Earth is used to power living systems.

4.61  T  **F**      Long-wave radiation easily passes through the atmosphere and escapes from Earth.

4.62  T  **F**      Atmospheric temperature increases with an increase in altitude.

4.63  **T**  F      The shape of Earth's orbit and the tilt of the planet's axis influences the concentration of solar         energy so that the sun's energy does not reach        Earth uniformly.

4.64  **T**  F    The law of tolerance supports the adage of
                 moderation in all things.

4.65  **T**  F    A positive feedback occurs when a change drives a
                 system to higher or lower values.

4.66  T  **F**    Turbulence in the atmosphere is caused by cold air
                 rising and warm air sinking.

4.67  **T**  F    During temperature inversions air pollution
                 increases.

4.68  T  **F**    The linear velocity of Earth increases toward the
                 poles and decreases at the equator.

4.69  **T**  F    The Coriolis effect deflects air and water
                 movements and prevents a simple flow from the
                 equator to the poles.

4.70  **T**  F    Intertropical convergences are the result of
                 changes in the altitude of the sun and distance
                 from the equator.

## MATCHING

Concerning the disposition of solar energy:

4.71  __**D**__    Incoming solar radiation

4.72  __**B**__    Reflected by the atmosphere

4.73  __**A**__    Absorbed by Earth's surface

4.74  __**C**__    Emitted by the atmosphere

4.75  __**B**__    Absorbed by the atmosphere

         A.  45%
         B.  25%
         C.  66%
         D.  100%

## DISCUSSION

4.76 Describe some possible effects atmospheric pollution    may
have on world climate.

4.77 What is the green house effect?  How do the events in the
    atmosphere differ from those in a glass greenhouse?

129

4.78 Climate conditions vary from one location to another.
     Discuss some conditions that may result in these local
     variations.

4.79 How do north-facing slopes differ from south-facing slopes?
     List the variations in microclimates on the two slopes and
     discuss how these variations influence the plant communities
     in each area.

4.80 What physical factors influence the global circulation of
     the atmosphere?  What influence do these factors have on the
     world-wide distribution of plants and animals?

# Chapter 5

## Temperature

### SENTENCE COMPLETION

5.1 **Conduction** is the movement of heat between a warm solid object and a cool one.

5.2 Heat transfer by **convection** will take place when either air or water moves over an object.

5.3 **Thermal radiation** results when solar energy is absorbed by an object and then is converted into heat.

5.4 Frogs and salamanders are classified as **poikilotherms** because their body temperature remains in equilibrium with the temperature of their environment.

5.5 **Active temperature range** is the range of body temperatures at which poikilotherms carry out their daily activities.

5.6 Fish can adjust to seasonal changes in water temperatures by **acclimatization**.

5.7 **Homoiotherms** maintain a constant body temperature despite environmental temperatures.

5.8 As body weight **increases** the basal metabolic rate decreases.

5.9 A **doubling** in body mass will increase the basal rate of metabolism by **75** percent.

5.10 **Fur** is a mammal's major form of insulation.

5.11 Hibernating mammals can increase heat production by burning **brown fat**.

5.12 **Thermogenesis** is the metabolism of brown fat to increase body heat.

5.13 Mammals increase evaporative cooling by **panting**.

5.14 Camels **store** body heat during the day and **release** this excess heat at night.

5.15 Arctic marine fish experience **supercooling** but do not freeze because of the presence of special chemicals in their body.

5.16 A dolphin prevents excessive heat loss from their fins by **countercurrent** circulation.

5.17 Hibernation is a **diapause** that occurs in winter.

5.18 Many shrubs turn sensitive cells into cold resistant ones by **frost hardening.**

5.19 All organisms have an **optimum** temperature at which they best maintain themselves.

5.20 **Anadromous** fishes spawn in fresh water and then migrate to salt water.


**MULTIPLE CHOICE**

5.21 The movement of heat by diffusion from a warm solid object to a cold one is called:
    a) convection;
    **b) conduction;**
    c) evaporation;
    d) thermal radiation.

5.22 Heat gain_____ heat loss.
    **a) equals;**
    b) is less than;
    c) is greater than;
    d) is sometimes greater than and sometimes less than.

5.23 Animals_____ heat by metabolism.
    a) lose;
    b) store;
    **c) gain;**
    d) radiate.

5.24 Poikilotherms:
    a) are warm-blooded;
    b) can regulate their internal body temperature;
    c) include mammals and birds;
    **d) have body temperatures that vary as environmental temperatures vary.**

5.25 Homoiotherms are:
    a) ectothermic;
    **b) endothermic;**
    c) both ectothermic and endothermic;
    d) ectothermic in winter and endothermic in summer.

5.26 A drop in environmental temperature will _____ the rates of enzymatic activities in poikilotherms.
    **a) decrease;**
    b) increase;
    c) have no effect on;
    d) first increase, then decrease.

5.27 Endotherms have an advantage over ectotherms in:
      a) tropical climates;
      **b) cold environments;**
      c) rainy environments;
      d) hot environments.

5.28 Which of the following is a temperature regulating mechanism found in endotherms?
      a) behavioral responses such as curling up into a ball;
      b) vasoconstriction of peripheral blood vessels;
      c) shivering;
      d) sweating;
      **e) all of the above are temperature regulating mechanisms.**

5.29 Small animals:
      a) have a low metabolic rate;
      b) require less food per unit of body weight than large animals;
      c) spend little time looking for and consuming food;
      **d) have a high metabolism.**

5.30 A rete is:
      **a) a net of intermingling blood vessels;**
      b) a sweat gland found in sea birds;
      c) a ratio of biomass to surface area;
      d) a type of blood fat used to insulate the bodies of arctic animals.

5.31 Which of the following body structures would help an animal get rid of excess body heat?
      a) subcutaneous fat;
      b) light-colored feathers;
      **c) long ears;**
      d) large body.

5.32 A camel is able to tolerate the high temperatures of the desert by:
      a) sweating and panting;
      b) storing water in its hump;
      c) burning brown fatty tissue;
      **d) allowing its body temperature to drop at night.**

5.33 Pigs, horses, and humans can correct a small rise in their body temperature by:
      **a) sweating;**
      b) releasing body heat at night;
      c) constricting the arteries of the skin;
      d) rapid contraction of striated muscles.

5.34 A drop in your body temperature would be corrected by:
  a) relaxing subcutaneous muscles;
  b) sweating;
  c) dilating the arteries of the skin;
  **d) shivering.**

5.35 Generally arteries carry:
  **a) warm blood;**
  b) cool blood;
  c) blood low in metabolic heat;
  d) blood low in glucose.

5.36 In countercirculation:
  **a) warm arterial blood exchanges heat with cool venous blood;**
  b) cool arterial blood exchanges heat with warm venous blood;
  c) warm arterial blood mixes with cool venous blood;
  d) cool arterial blood mixes with warm venous blood.

5.37 The amount of heat that moves into or out of a body is proportional to the body's _____.
  a) volume;
  b) mass;
  c) weight;
  **d) surface area.**

5.38 The internal temperature of a plant is maintained by:
  a) its metabolism;
  b) fat deposits in its cuticle;
  c) supercooling of its sap;
  **d) the position of its leaves and the reflectivity of its bark and leaves.**

5.39 A mammal insulates its body against heat loss by:
  a) light-colored fur;
  b) scaled tarsi;
  c) countercurrent circulation;
  **d) subcutaneous fat and hair.**

5.40 Changes in the temperature of a river may:
  a) interfere with the migration of fish;
  b) result in cool water diatoms being replaced by green algae;
  c) lead to a decline in the number of carp and bass;
  **d) a,b, but not c;**
  e) a,b, and c.

**TRUE/FALSE**

5.41  T  **F**  Heat transfer by conduction is faster than by convection.

5.42  **T**  F  Endotherms depend on internal physiological mechanisms to maintain their body temperature.

5.43  T  **F**  Poikilotherms have a high metabolic rate and rarely conduct heat between their body and the environment.

5.44  T  F  Ectotherms can slow down metabolic activities at times of temperature extremes.

5.45  **T**  F  Ecologist believe that high body temperatures evolved form the inability of animals to get rid of the heat they produced during periods of high metabolic activity.

5.46  **T**  F  Homoiotherms have a high metabolic rate, yet they do not lose heat easily to the environment.

5.47  **T**  F  Small animals require more food per unit of body weight than larger animals.

5.48  T  **F**  Large animals require more food per unit of body weight than smaller animals.

5.49  T  **F**  Large animals lose more heat to the environment than smaller ones.

5.50  **T**  F  Young homoiotherms begin their life as ectotherms.

5.51  **T**  F  Small body size is more likely the case for ectotherms than for homoiotherms.

5.52  **T**  F  If environmental temperatures fall to a point beyond which body insulation can no longer prevent body heat loss, homoiotherms may increase their metabolic rate.

5.53  T  **F**  Blood entering the flipper of an arctic porpoise is warmer than the blood leaving the flipper.

5.54  T  **F**  A dormancy experienced by some desert animals during the summer is called hibernation.

5.55  T  **F**  Bears hibernate in winter to reduce their metabolic activity.

5.56  **T**  F    The metabolic activity of plants does not contribute to their internal temperature.

5.57 **T**  F    Plants growing in shady environments have leaves arranged perpendicular to the sun's rays.

5.58  T  **F**    Plants with deeply lobed leaves conserve heat very effectively.

5.59  **T**  F    A temperatures above 55 degrees Celsius the nucleic acids in plant cells can be damaged.

5.60  **T**  F    Some plants, such as skunk cabbage, <u>Symplocarpus foetidus,</u> are endothermic for a least short periods of time.

## MATCHING

    A.  conduction
    B.  evaporation
    C.  radiation
    D.  convection

5.61  __**A**__    A dog digs a hole in the cool earth and spreads out in order to lose heat.

5.62  __**B**__    An overheated crane urinates on its legs in order to lose heat.

5.63  __**D**__    On a hot day a turkey vulture sits on top of a fence post, where there is a breeze, in order to lose heat.

5.64  __**C**__    A snake basks in the sun in order to acquire heat.

5.65  __**A**__    When a walrus' foot contacts the ice it loses heat by____.

    A.  poikilotherm
    B.  homeotherm
    C.  endotherm
    D.  ectotherm

5.66.__**D**__    An animal that warms its body by absorbing heat from its environment.

5.67  __**B**__    "Warm-blooded" animals.

5.68  __**C**__    Skunk cabbage, philodendrons and some members of the arum lily family.

5.69  __**A**__    An animal whose body temperature fluctuates with the environmental temperature.

## DISCUSSION

5.70    What effect would human-induced global warming have on the vegetation of Earth?

5.71    If ectotherms lack physiological controls of their metabolic rate, how can they regulate their body temperature?

5.72    Heat moves from warmer areas to colder areas. How can an animal lose body heat when its environment is warmer than its body?

5.73    How does an endotherm lower its body temperature when it is too high?

5.74    What happens to an endotherm during hypothermia?

5.75    How do plants control their internal body temperature?

# Chapter 6

## Moisture

SENTENCE COMPLETION

6.1 **Covalence** is the sharing of electrons.

6.2 In a water molecule each **hydrogen** proton shares its electron with **oxygen**.

6.3 Water molecules are **electropositive** on the side with the hydrogen atoms and electronegative on the opposite side.

6.4 Water molecules are joined by **hydrogen bonds**.

6.5 The **specific heat** of water is the amount of heat that is required to raise the temperature of water one degree Celsius.

6.6 It takes **80** calories of heat to convert **one** gram of ice at **one** degree Celsius to a liquid state.

6.7 Water behaves as though it consists of a series of parallel concentric layers. The resistance between the layers is called **viscosity**.

6.8 The attraction of water molecules to other water molecules results in **surface tension**.

6.9 Water **absorption** in organisms must equal water evaporation.

6.10 **Evaporation** is the physical process of converting liquid water to vapor.

6.11 **Transpiration** in plants is driven by the evaporation of water at the leaf/atmosphere interface.

6.12 Some desert plants have seeds that remain dormant waiting for the right amount of moisture to stimulate them. This is an example of **drought avoidance**.

6.13 **Xeric** species of plants are adapted to dry conditions.

6.14 Plants that can store water in cells are known as **succulents**.

6.15 **Phreatophytes** have long tap roots that allow the plants to use water deep beneath the ground's surface.

6.16 Oceans, salt marshes and alkaline deserts are **saline** environments.

6.17 Pumping salts against a **concentration** gradient is one type of **active transport.**

6.18 Salt marsh plants are called **halophytes.**

6.19 **Climographs** are composite pictures of mean monthly temperatures and relative humidities of an area.

6.20 **Rain shadows** result in dry desert-like conditions on the leeward slopes of mountains.

**MULTIPLE CHOICE**

6.21 When water freezes the arrangement of water molecules is:
　　　　a) **regular and open;**
　　　　b) regular and condensed;
　　　　c) random and open;
　　　　d) random and condensed.

6.22 Large bodies of water maintain a  constant temperature because of:
　　　　a) **water's high specific heat;**
　　　　b) water's high viscosity;
　　　　c) water's high surface tension;
　　　　d) water's covalent nature.

6.23 In order for organisms to maintain a water balance:
　　　　a) water taken up by an organism must be greater than the water lost;
　　　　b) water taken up by an organism must be less than the water lost;
　　　　c **water taken up by an organism must be equal to the water lost;**
　　　　d) the rate of absorption must be less than the rate of transpiration.

6.24 _____is the physical process of converting liquid to water vapor.
　　　　a)　 **Evaporation;**
　　　　b)　 Drought tolerance;
　　　　c)　 Transpiration;
　　　　d)　 Evapotranspiration.

6.25 Which of the following plants has little or no protection against water loss?
　　　　a) pine trees;
　　　　b) cacti;
　　　　c) **algae;**
　　　　d) flowering plants.

6.26 Xeric species are adapted to:
    a)  cold conditions;
    b)  wet conditions;
    **c)  dry conditions;**
    d)  high altitude.

6.27 Which of the following is not a plant response to drought?
    **a)  open stomata;**
    b)  transpire through cuticle;
    c)  store carbon for photosynthesis;
    d)  change the structure of lipids in the cuticle.

6.28 Fresh water organisms:
    a)  must conserve water;
    b)  must conserve energy;
    c)  must get rid of excess salts;
    **d)  must get rid of excess water.**

6.29 You would most likely find salt glands in:
    a)  freshwater fish;
    b)  freshwater ducks;
    c)  desert mammals;
    **d)  saltwater birds.**

6.30 The windward side of a mountain would support:
    a)  xeric species;
    b)  phreatophytes;
    **c)  mesic species;**
    d)  succulents and shrubs.

## TRUE/FALSE

6.31  **T**  F    The lower the temperature of water the greater its density.

6.32  **T**  F    Ammonia is one substance with a higher specific heat than water.

6.33  T  **F**    Because of the nature of water, lakes tend to cool off rapidly in the fall and warm up quickly in the spring.

6.34  T  **F**    Animals expend more muscular energy moving through air than through water.

6.35  **T**  F    An animal with a short, rounded head and a tapering body is perfectly designed for moving in water.

6.36  T  **F**    Drought, the most common form of water stress on plants, is not a problem during winter months.

6.37  **T**  F  Drought resistance is dependent on a plant's drought tolerance and drought avoidance.

6.38  **T**  F  Cacti lack leaves but can carry on photosynthesis in their stems.

6.39  **T**  F  Estivation is one way small animals can avoid water loss.

6.40  **T**  F  A saline environment is very similar to a desert.

6.41  T  **F**  Halophytes have very low concentrations of ions within their cells.

6.42  T  **F**  Halophytes grow best on nonsaline soil.

6.43  **T**  F  The Pacific Northwest supports coniferous rain forests because the climate is mild and wet.

6.44  **T**  F  Rainfall is heavier on the windward side of a mountain than on the leeward side.

6.45  T  **F**  Many arctic animals produce highly concentrated urine and feces.

6.46  T  **F**  Water moves in the direction of high water potential.

**MATCHING**

    A.  surface tension
    B.  specific heat
    C.  heat of evaporation
    D.  viscosity

6.47  __**A**__  Enables some organism to move across the surface water.

6.48  __**B**__  Important in temperature regulation of bodies of water.

6.49  __**C**__  Effective cooling mechanism for body temperature regulation

6.50  __**D**__  Forces aquatic animals to expend considerable muscular energy moving through water.

**DISCUSSION**

6.51      Why does ice float in water?

6.52      Why is the evaporation of perspiration an effective
mechanism for regulating body temperature?

6.53      What physiological adaptations would you expect to find
in fish that live in fresh water?  How would these
adaptations differ in fish living in saltwater?

6.54      What special problems do saline environments present to
plants and animals that live there?  List
some of the special adaptations to saline environments
you would expect to find in the plants and animals.

6.55      How do mountains modify the regional vegetation?  How
do rain shadows form?

# Chapter 7

## Light

<u>**SENTENCE COMPLETION**</u>

7.1  **Photosynthetically active radiation** consists of energy wavelengths between 0.40 microns and 0.70 microns.

7.2  Leaves reflect **green** light.

7.3  Chloroplasts absorb **red** light and use it in photosynthesis.

7.4  **Shade-tolerant** plants have a lower rate of photosynthesis and a lower rate of respiration than shade avoiders.

7.5  Very little photosynthesis occurs in water depths below **100** meters.

7.6  A plant requires **carbon dioxide** and **water** to manufacture carbohydrates.

7.7  The two stages of photosynthesis include the **light reaction** and the **dark reaction**.

7.8  Light energy is stored in high-energy phosphate bonds in molecules of **ATP**.

7.9  **RuBP** combines with carbon dioxide to produce an intermediate six-carbon molecule in the Calvin cycle.

7.10 **Photorespiration** may be a way for plants to release energy when carbon dioxide is not available.

7.11 Carbon dioxide is converted into sugars and starches during the **dark reaction** of photosynthesis.

7.12 **CAM** plants fix carbon dioxide from the atmosphere as well as from malate.

7.13 The intermediate compound in photosynthesis is **glucose**.

7.14 Reactions to **moisture** as well as to light intensity will decide whether a plant is shade tolerant or not.

7.15 **Foliage density** is expressed as leaf area index.

7.16 What factor is not important in limiting the rate of
     photosynthesis?
     a) light;
     b) water;
     **c) oxygen;**
     d) carbon dioxide.

7.17 To allow sufficient carbon dioxide to enter a leaf, plants
     keep their_____open.
     a) vascular bundles;
     **b) stomata;**
     c) mesophyll;
     d) bundle sheath cells.

7.18 The high level ozone layer absorbs _____
     wavelengths of light.
     a) red;
     b) blue;
     c) green
     d) violet;
     **e) all.**

7.19 Leaves appear green because they reflect_____
     wavelengths of light.
     **a) green;**
     b) blue;
     c) red;
     d) orange.

7.20 Chloroplasts in the mesophyll of a leaf absorb_____light
     and use it in photosynthesis.
     a) green;
     b) yellow;
     **c) red;**
     d) violet.

7.21 In a meadow the greatest absorption of photosynthetically
     active radiation (PAR) occurs at:
     a) the upper crown of vegetation;
     b) the lower crown of vegetation;
     **c) the middle and lower regions of vegetation;**
     d) the ground.

7.22 Young plant shoots and rapidly developing leaves synthesize:
     a) simple sugars;
     b) nucleic acids;
     **c) fats and proteins;**
     d) free amino acids.

7.23 Dry weather will cause a plant to:
   a) lose water;
   b) close its stomata;
   c) shut down its C3 cycle;
   d) a and b, but not c;
   **e) a, b, and c.**

7.24 The 4-carbon pathway tolerates low _____ concentrations.
   **a) carbon dioxide;**
   b) oxygen;
   c) nitrogen;
   d) water.

7.25 During the light reaction of photosynthesis:
   a) carbon dioxide os fixed into carbohydrates;
   **b) light energy is converted into chemical energy and is passed on to ATP and NADPH;**
   c) RuBP combines with carbon dioxide;
   d) ATP is used to fix carbon dioxide.

7.26 During the light reaction of photosynthesis _____ is active at 700 nanometers and_____ is active at 680 nanometers.
   **a) photosystem I, photosystem II;**
   b) photosystem II, photosystem I;
   c) ATP, NADPH;
   d) ADP, NADP.

7.27 In photorespiration _____ not carbon dioxide is fixed to RuBP.
   a) nitrogen;
   b) hydrogen;
   **c) oxygen;**
   d) ozone.

7.28 CAM plants are most likely to be found in:
   a) forests;
   **b) deserts;**
   c) tundra;
   d) grasslands.

7.29 C4 plants probably evolved in_____climates.
   **a) tropical;**
   b) temperate;
   c) arctic;
   d) cold.

7.30 C4 plants use more ATP than C3 plants, but the C4 plants can survive where _____ is limited in abundance.
         **a) water;**
         b) light;
         c) oxygen;
         d) RuBP.

## TRUE/FALSE

7.31   T   **F**       The ozone layer of the atmosphere absorbs the red and orange wavelengths of light.

7.32   T   **F**       The sky appears blue because the blue wavelengths of light are absorbed by gases in the atmosphere.

7.33   **T**   F       Leaves arranged in a horizontal direction trap more sunlight than leaves arranged in an upright position.

7.34   **T**   F       Shade avoiders will be the first plants to colonize a disturbed area.

7.35   **T**   F       Shade avoiders grow rapidly.

7.36   T   **F**       Aquatic habitats support plants that are adapted to growing in bright light.

7.37   **T**   F       Except for a small group of bacteria photosynthesis is restricted to plants which contain chlorophyll in their cells.

7.38   **T**   F       The light reaction of photosynthesis converts light energy into chemical energy which is passed to the energy carriers ATP and NADPH.

7.39   T   **F**       During the dark reaction of photosynthesis ATP and NADPH are produced.

7.40   T   **F**       When plants close their stomata in hot dry weather the C4 cycle shuts down due to a deficiency of carbon dioxide.

7.41   **T**   F       Factors limiting the rate of photosynthesis are the amount of light, water and carbon dioxide.

7.42   **T**   F       C4 pathway of photosynthesis gives the plants that possess it an advantage in hot and dry climates.

7.43    T   **F**       CAM plants open their stomata during the day to absorb carbon dioxide and then close them at night to prevent a carbon dioxide loss.

7.44  **T**  F    It is possible that C4 plants may be more
                  competitive and more adaptable than C3 plants.

7.45  **T**  F    When you consider the energy input required to
                  begin the process, photosynthesis is not very
                  efficient.


**MATCHING**

A.  C3 plants
B.  C4 plants

7.46  __**A**__    Using sunlight, chlorophyll splits water.

7.47  __**A**__    ATP and NADPH are produced.

7.48  __**A**__    Photosystems absorb solar radiation.

7.49  __**B**__    RuBP accepts molecules of carbon dioxide.

7.50  __**B**__    The Calvin-Benson cycle will lead to the
                   production of glucose.

**DISCUSSION**

7.51    What specific problems does a hot, dry environment
        create for a plant?

7.52    If you were growing plants under artificial light what
        wavelengths of light must be present in order for
        photosynthesis to take place?  What wavelengths of
        light are unnecessary for photosynthesis?

7.53    Explain the main steps in the light reaction and the
        dark reaction. How are the two reactions coupled
        together during photosynthesis?

7.54    Compare the mechanisms of C3 and C4 photosynthesis.

7.55    "Photosynthesis is a rather inefficient process".  Is
        there evidence to support this statement?

# Chapter 8

## Periodicity

8.1 The recurrence of daily and seasonal changes are called **rhythmicity.**

8.2 **Photoperiodism** is a negative response to changes in light intensity (light and darkness).

8.3 In most vertebrates metabolic rate, temperature, and general level of nervous activity alter regularly in a 24-hour cycle. Such a cycle is called a **Circadian rhythm.**

8.4 The length of a circadian rhythm is a self-sustained oscillation; it is **free-running.**

8.5 Environmental cues or synchronizers in circadian rhythms are called **zeitgebers.**

8.6 **Jet lag** results when the master clock is reset but subservient clocks are out of phase for a time.

8.7 **Entrainment** refers to the fact that the environment cue, light, brings the circadian rhythm of many organisms into phase with the 24-hour photoperiod.

8.8 A change in the time an of organism's response to an environmental cue that will alter general activity is called a **phase shift.**

8.9 Light-requiring cycles are **photophilic** and dark-requiring cycles are **scotophilic.**

8.10 In birds the biological clock is located in the **pineal gland.**

8.11 **Critical daylength** defines a period of light that when reached either inhibits or promotes a response.

8.12 Organisms not affected by the length of day or night are **day neutral.**

8.13 Plants that will flower when they are exposed to daylengths shorter than the critical daylength are called **short-day plants.**

8.14 A time of arrested growth, generally during the winter is called **diapause.**

148

8.15 **Long-day plants** are stimulated to flower when daylength increases in late spring and summer.

8.16 In birds the time when light cannot induce gonadal development is known as the **refractory period.**

8.17 Photoperiod can stimulate **seasonal morphs**, distinct changes in the animal's phenotype during a year.

8.18 Animals that hibernate and most birds seem to respond to a 365-day cycle or a **circannual** clock.

8.19 The change from day to night also will be accompanied by a(n)_____ **increase**/decrease in humidity and a(n) _____increase/**decrease** in temperature.

8.20 The synchronization of the behavior of a population to environmental cues is the result of **natural selection.**

## MULTIPLE CHOICE

8.21 Which of the following is not an example of rhythmicity?
      a) birds sing as dawn breaks;
      b) nocturnal animals are rarely active during daytime;
      **c) a lizard loses its tail when attacked by a predator;**
      d) as winter approaches animals begin to store fat.

8.22 Daily periodicity, an activity rhythm with a 24-hour day includes:
      a) pattern of leaf and petal movement;
      b) sleep pattern in animals;
      c) emergence of insects from pupal cases;
      d) a & b, but not c;
      **e) a, b & c.**

8.23 Biological rhythms respond to:
      a) light intensity;
      b) temperature;
      c) exogenous stimuli;
      **d) exogenous and endogenous stimuli.**

8.24 If a nocturnal animal that responds to photoperiod is kept in total darkness:
      **a) it will still maintain a constant rhythm of activity from day to day;**
      b) it will not be able to maintain a rhythm of daily activity because it does not experience daylight;
      c) it first drifts out of phase with the day-night change, then it resets its biological clock;
      d) its activity level drops considerably and it begin to hibernate.

149

8.25  Which of the following is an example of a biological clock
      but not a circadian rhythm?
      a) leaves of some plants are oriented horizontally
         during daylight but take a compact vertical
         position at night;
      b) human body temperature fluctuates by 2 degrees
         Fahrenheit during the day, reaching a
         high at 4:00 PM and a low at 4:00 AM;
      c) flying east from America to Europe a traveler may
         have several days of difficulty in reestablishing
         normal waking-sleeping patterns;
      **d) in the South Pacific a small marine worm comes to
         the surface of the sand once every November.**

8.26  Zeitgeber is:
      a) the number of hours in a rhythm from the beginning of
         activity on one day to the beginning of activity on
         the next;
      b) the response of an organism to changing light and
         darkness.
      **c) an external cue such as temperature and light that
         synchronizes the activity of an organism with the
         environment;**
      d) the period of greatest gonadal development in the
         reproductive cycle of birds.

8.27  One of the most important synchronizers for circadian
      rhythms is:
      a) rainfall;
      b) altitude;
      c) soil type;
      **d) temperature.**

8.28  In multicellular animals the biological clock is most
      likely associated with the:
      **a) brain;**
      b) reproductive organs;
      c) circulatory system;
      d) endocrine glands.

8.29  In mammals the biological clock is located in the:
      a) optic-nerve;
      b) pineal gland;
      **c) suprachiasmatic nuclei;**
      d) optic lobes.

8.30  In birds the biological clock is located in the:
      a) optic-nerve;
      **b) pineal gland;**
      c) suprachiasmatic nuclei;
      d) optic lobes.

8.31 The biological clock:
a) should have a 48-hour cycle;
b) cannot be changed by changes in the length of day;
c) requires some type of environmental timesetter to make it run continuously;
**d) must be able to run the same at all temperatures.**

8.32 The first half of a circadian rhythm of sensitivity to light is the:
**a) photophilic phase;**
b) photophobic phase;
c) scotophilic phase;
d) scotophobic phase.

8.33 The last half of a circadian rhythm of sensitivity to light is the:
a) photophilic phase;
b) photophobic phase;
**c) scotophilic phase;**
d) scotophobic phase.

8.34 The biological clock:
a) creates a sense of time for plants and animals;
b) may create a time memory for some animals;
c) helps some animals to orient themselves to the sun;
d) helps animals to prepare for major seasonal events;
**e) all the above describe biological clocks.**

8.35 Crabs and other tidal animals that are brought into the laboratory and held under constant conditions of temperature and light:
a) lose their rhythms to tides;
b) will respond to the times the tides would normally ebb, but not to the times when the tides would flow;
**c) exhibit the same rhythms to tides as they would back in their natural environment;**
d) a, b, but not c;
e) a, b, and c.

8.36 Critical day length in most organisms is between:
a) 3 and 6 hours;
**b) 10 and 14 hours;**
c) 20 and 24 hours;
d) 18 and 20 hours.

8.37 Plants and animals that are not affected by daylength are:
**a) day neutral organisms;**
b) short-day organisms;
c) long-day organisms;
d) photophobic organisms.

8.38  T  **F**  Scientists believe that organisms respond only to exogenous stimuli such as light intensity, temperature, and tidal changes.

8.39  T  **F**  If nocturnal animals are exposed to continuous light they will become active for as long as they are exposed to the light.

8.40  **T**  F  The "master" synchronizer for circadian rhythms is light.

8.41  T  **F**  It is impossible to reset a circadian rhythm to another cycle.

8.42  **T**  F  Circadian rhythms allow an animal to anticipate regular daily events before an environmental cue appears.

8.43  T  **F**  When intertidal animals are brought into the laboratory and held under constant conditions lose their response to tidal cycles.

8.44  T  **F**  In birds the biological clock is located in the suprachiasmatic nuclei.

8.45  **T**  F  Generally biological clocks have 24-hour rhythms.

8.46  T  **F**  Biological clocks can be disrupted by changes in the environmental temperature.

8.47  **T**  F  According to the Bunning model the first half of a rhythm to light is photophilic and the second is scotophilic.

8.48  **T**  F  Most organisms have a critical daylength of 10 to 14 hours.

8.49  **T**  F  Plants that bloom during the winter months are long-day plants.

8.50  **T**  F  Some insects that exhibit diapause react to the lengthening days of spring by resuming larval development.

8.51  **T**  F  After breeding the reproductive organs of birds shrink in size.

8.52  T  **F**    During the progressive phase of a bird's reproductive cycle exposing the animal to a long-day photoperiod will induce gonadal atrophy.

8.53  T  **F**    The fact that some animals have two coat colors, a light one in winter and a dark one in summer is an example of diapause.

8.54  **T**  F    As the daylight hours increases and the night hours decrease the testes of some male animals will descend into the scrotum.

8.55  T  **F**    Antler growth in deer is triggered by the shortening days of winter.

8.56  T  **F**    In order for a circannual rhythm to exist, the rhythm must respond to changes in temperature.

8.57  **T**  F    Predators often synchronize their feeding behavior to the activity rhythms of their prey.


**MATCHING**

    A.  daily cycle
    B.  lunar cycle
    C.  tidal cycle
    D.  annual cycle

8.58  __D__    The arctic fox undergoes a seasonal change in its coat color.

8.59  __A__    Bison feed only during daylight hours and sleep when it gets dark.

8.60  __B__    Human females experience an ovulation once every 28 days.

8.61  __C__    Shorebirds move into mudflats to feed as the tidal waters recede.

    A.  entrainment
    B.  zeitgeber
    C.  master clock
    D.  free-running
    E.  phase shift

8.62  __A__    A process that is analogous to setting a watch at the correct time.

8.63 __B__    An environmental stimulus that sets a biological clock.

8.64 __E__    A series of stepwise delays that adjust an organism's biological rhythm.

8.65 __D__    The period of a circadian rhythm.

8.66 __C__    The highest level in a hierarchy of oscillators.


## DISCUSSION

8.67    Are biological clocks regulated by internal rhythms or are they in tune with some external factor?

8.68    What is a biological clock?  Give an example of one that is cycled annually, one that is cycled monthly, and one that is cycled daily.

8.69    The accident at Three Mile Island nuclear power plant occurred at 4:00 A.M. with a crew on duty that had just rotated to the night shift.  Could there be a biological explanation for this accident?  How would you design a change of work shift to modify the possible negative effects of changes in time?

8.70    What are some adaptive values of circadian rhythms?

154

# CHAPTER 9

## SOIL

### SENTENCE COMPLETION

9.1 Loose material transported from one area to another is called **alluvial.**

9.2 **Till** refers to material transported by glacial ice.

9.3 Soil materials which come from organic matter which are not carried from one place to another are called **residual.**

9.4 A cross-section of a soil horizon is called a **soil-profile.**

9.5 **Humus** is a complex mixture of partially decomposed organic matter mixed with various inorganic compounds.

9.6 Distinct layers in the soil created by localized chemical and physical processes are called **horizons.**

9.7 There are four major horizons in soil: the **O** horizon, **A** horizon, **B** horizon, and the **C** horizon. The **R** horizon is the layer of bedrock below the other horizons.

9.8 From an ecological point of view the **R** horizon is the most important part of a soil profile.

9.9 Soils that develop by **calcification** have a thick A horizon and a B horizon in which **calcium** compounds have accumulated.

9.10 **Podzol** soils are characterized by B horizon rich in iron and aluminum compounds and an A horizon fireplace ash; sometimes a cement like layer called a hard pair forms below the B horizon.

9.11 In tropical regions of the world, heavy rainfall and high temperature develops a unique soil type formed by the soil process called **laterization.**

9.12 In poorly drained soils a process called **gleization** can result a soil with compact horizons, rich in organic matter and free of soil microorganisms.

9.13 The number of negatively charged sites on a soil particle that can attract positively charged cations is called the **cation exchange capacity.**

9.14 **Micelles** are platelike particles in the soil. The interior of these plates are electrically balanced but the edges of the plates carry a negative electrical charge.

9.15 **pH** is used as a measure of the degree of acidity or alkalinity of a soil.

9.16 **Soil texture** is determined by the different sizes and types of particles in soils.

9.17 **Clay particles**, which are colloidal in nature are too small to be seen under a light microscope.

9.18 If the pore spaces in soils are filled with water and there is an excess of water draining out of the soil, the soil is **saturated.**

9.19 **Serpentine** soils are low in calcium and high in other elements such as nickel and magnesium.

9.20 **Soil erosion** is the movement of the components of soil from one area to the next. The movement generally involves the displacement of topsoil.

## MULTIPLE CHOICE

9.21 Broken rock transported from one place to another by wind is known as:
> **a) loess;**
> b) till;
> c) alluvial;
> d) residual.

9.22 The primordial mass of material from which different types of soil can develop is called the:
> a) mantle;
> **b) regolith;**
> c) lacrustine;
> d) till.

9.23 The layers that soils become organized into are known as:
> a) profiles;
> b) tills;
> c) negoliths;
> **d) horizons.**

9.24 The organic matter is found in the ___ horizon.
> a) O;
> **b) A;**
> c) B;
> d) C.

9.25 Which of the following horizons would contain no organic litter?
        a) A;
        **b) C;**
        c) O;
        d) B.

9.26 The zone of maximum biological activity is the ___ horizon:
        **a) A;**
        b) B;
        c) C;
        d) O.

9.27 Humus is:
        a) partially decomposed organic matter;
        b) generally dark brown or black in color;
        c) the home for many microorganisms;
        **d) all of the above.**

9.28 Minerals tend to accumulate in the ___ horizon:
        a) A;
        b) B;
        **c) C;**
        d) O;
        e) R.

9.29 The biological or chemical breakdown of the regolith into soil is called:
        a) leaching;
        b) mineralization;
        c) percolation;
        **d) weathering.**

9.30 The soils in tropical rain forests are characterized as:
        a) aridisols;
        b) spodosols;
        c) oxisols;
        **d) mollisols.**

9.31 There are ___ major soil orders.
        a) three;
        b) twenty;
        **c) eleven;**
        d) five.

9.32 Leaching takes place in soils when:
        a) humus is decomposed by bacteria;
        b) salts are deposited as surface water evaporates;
        c) soil water accumulates and fails to drain away;
        **d) water removes soluble nutrients.**

9.33 The texture of a soil is determined by the proportion of which of the following soil particles?
    a) clay, sand, loam;
    **b) clay, silt, sand;**
    c) silt, humus, loam;
    d) clay, sand, humus.

9.34 ___ controls the most important properties of soil.
    **a) clay;**
    b) silt;
    c) sand;
    d) humus.

9.35 Soils with high water content are also deficient in:
    a) magnesium;
    b) iron;
    **c) oxygen;**
    d) clay.

9.36 Podzol soils may develop in:
    a) tropical rain forests;
    b) grasslands;
    **c) temperate forests;**
    d) deserts.

9.37 You would expect that agriculture crops would grow well on ___ soil type.
    a) oxisol;
    b) vertisol;
    **c) mollisol;**
    d) amdisol.

9.38 Which of the following was not responsible for the 1930 Dust Bowl?
    **a) crop rotation;**
    b) drought;
    c) overgrazing;
    d) grassland fires.

9.39 Which of the following is likely to cause soil erosion?
    a) off-road vehicles;
    b) logging;
    c) soil compaction;
    d) a, b, but not c;
    **e) a, b, + c.**

9.40 Which continent has suffered extensive desertification?
    a) North America;
    b) Asia;
    **c) Africa;**
    d) Australia.

**TRUE/FALSE**

9.41　T　**F**　Soil can be defined as the abiotic environment for plants formed from weathered rocks.

9.42　**T**　F　Small pieces of rock transported water from one area to another form alluvial deposits.

9.43　**T**　F　The mantle of unconsolidated rock upon which soil is formed is called the regolith.

9.44　**T**　F　The first steps in the alteration of deposited material and the formation of soil begins with the activity of some type of plant growth.

9.45　T　**F**　New humus is constantly being formed by the process of mineralization.

9.46　**T**　F　Heavy rainfall results in heavy soil leaching and rapid chemical weathering.

9.47　**T**　F　The soil in a particular area is arranged in layers called horizon that will vary in texture structure and consistency from one another.

9.48　T　**F**　In a soil profile, horizon O is made up of the bedrock.

9.49　T　**F**　From an ecological standpoint the B horizon is the most important because it is the organic horizon and the site of all biological decomposition.

9.50　**T**　F　Since mull humus is acidic in nature few soil organisms except for a small group of fungi decompose it.

9.51　T　**F**　Soil developed by calcification with a thick A horizon and a B horizon contain large amounts of calcium carbonate is typical of forest ecosystems.

9.52　**T**　F　Organic matter tends to accumulate in forests and only a small portion is decomposed each year.

9.53　T　**F**　Forest soils are generally alkaline in nature.

9.54　**T**　F　High temperatures and heavy rainfall will result in a soil composed of silicate and hydrous oxides, low in plant nutrients, and having distinct horizons.

9.55 **T** F  The soil forming process called salinization generally occurs in areas with light precipitation.

9.56 T **F**  Gley soils are low in organic matter because they are rich in organic microorganisms that decompose the organic material as fast as it is formed.

9.57 T **F**  In soils ions are dissolved in water and their movement will obey the laws of diffusion.

9.58 **T** F  Plants growing in limestone soils have to be adapted to xeric conditions.

9.59 **T** F  Clay particles control many important properties of soil including its plasticity and the exchange of ions between soil solution and soil particles.

9.60 **T** F  Soil is an ideal medium for life because it is chemically and structurally stable.

## MATCHING

    A. O horizon
    B. A horizon
    C. B horizon
    D. C horizon
    E. R horizon

9.61 __**C**__  Minerals accumulate in this layer.

9.62 __**E**__  Bedrock.

9.63 __**D**__  Contains little or no organic matter.

9.64 __**A**__  Made up of leaf litter and organic deposits.

9.65 __**A**__  The topsoil.

## DISCUSSION

9.66 Define soil. Why is it difficult to develop a good definition?

9.67 How does physical, biological, and chemical weathering lead to the formation of soils?

9.68 How does the proportion if clay, silt, and sand determine the texture of soil?

9.69 What characteristics of soil make it a suitable environment for life?

9.70 How does overgrazing, deforestation, and poor land use patterns lead to soil erosion?

# Chapter 10

## Density, Distribution, and Age

**SENTENCE COMPLETION**

10.1      A **population** is a group of similar interacting
          organisms that find themselves in the same place
          simultaneously.

10.2      A single plant with its own genetic characteristics is
          termed a **genet**, but if the plant reproduces sub-
          populations by forming roots or suckers these
          clones are called **ramets**.

10.3      Three characteristics of all populations are **density**,
          **distribution** and **age structure**.

10.4      The density of a population may control its **birthrate**,
          **mortality rate**, and **growth**.

10.5      A population of mice is found at a density of 500
          mice/hectare.  This is a measure of the population's
          **crude density**.

10.6      The patterns of distribution of organisms over space
          may be **uniform**, **random**, or **clumped**.

10.7      There are 250 people living in my apartment house or
          3.2  people per apartment.  The first number is
          **abundance** and the second number is **density**.

10.8      The age structure of a population can be divided into
          three ecological periods: **pre-reproductive**,
          **reproductive**, and **post-reproductive**.

10.9      All continuously breeding populations exhibit a **stable
          age structure**, but to achieve this structure age-
          specific birthrates and mortality rates must remain
          **constant**.

10.10     If a population has a fixed proportion of
          individuals in its age classes it is exhibiting a
          **stationary age distribution**.

10.11     Whenever a population experiences an **age ratio shift**
          because  of changes in age-specific death rates the
          birthrate is affected.

10.12    The **age structure** of a population represents the ratio
         of the age classes in the population to each other at a
         given time.

10.13    In an age structure diagram the **pre-reproductive** age
         group is at the bottom of the diagram.

10.14    The age structure diagram of India shows a very large
         **pre-reproductive** age group.

10.15    When considering the age of a tree within a forest the
         **greater** the diameter of the tree, the **older** it is.


## MULTIPLE CHOICE

10.16    A group of similar, interacting individuals of the same
         kind living in the same place simultaneously is
         called a:
                   a) cline;
                   b) deme;
                   **c) population;**
                   d) genet.

10.17    Which of the following is not an attribute shared by
         all populations?
                   a) density;
                   b) distribution;
                   c) age structure;
                   **d) biomass.**

10.18    Which of the following is not an important factor when
         deciding crude density?
                   a) the number of organisms in the
                      population;
                   b) the space occupied by the population;
                   **c) the amount of area available as suitable
                      living space;**
                   d) all of these are important in determining
                      crude density.

10.19    If the position of individuals within a population is
         independent of each other, the distribution is said to
         be:
                   **a) random;**
                   b) clumped;
                   c) contagious;
                   d) uniform.

10.20    The most common type of distribution if individuals
         within a population is:
                    a) random;
                    **b) clumped;**
                    c) uniform;
                    d) continuous.

10.21    _____ is the number of individuals in a
         given area.
                    a) ecological density;
                    b) crude density;
                    c) range;
                    **d) abundance.**

10.22    Which of the following would result on temporal
         distribution?
                    a) circadian rhythms;
                    b) lunar rhythms;
                    c) tidal rhythms;
                    **d) all of the above.**

10.23    Direct count of all the individuals within a
         populations:
                    a) can be determined when doing density
                       studies, but cannot be used to find
                       abundance;
                    b) can be used to find species abundance
                       but not density;
                    c) can be best accomplished in populations
                       that exhibit a random distribution;
                    **d) are  time consuming or  expensive to
                       complete.**

10.24    Sampling is most useful in the study of:
                    a) large grazing animals;
                    b) predators;
                    c) insects;
                    **d) plants.**

10.25    When making an accurate estimate of crude density:
                    a) you also must know the abundance of a
                       population;
                    b) you must take a direct sample of the
                       entire population;
                    **c) you must know how individuals within the
                       population are distributed;**
                    d) you must know the age structure of the
                       population.

10.26     The age structure of a population is its:
                    a) relative density in a given area;
                    b) relative number of births per year;
                    c) relative number of deaths per year;
                    **d) relative number of individuals at each**
                    **age.**

10.27     Among annual grasses, the_____ period has little
          effect on the potential rate of population growth.
                    a) post-reproductive;
                    b) reproductive;
                    **c) pre-reproductive.**

10.28     The age pyramid for Sweden, a country approaching zero
          growth shows:
                    **a) a constant age structure;**
                    b) a large post-reproductive age class and a
                       small reproductive age class;
                    c) small reproductive age class and a large
                       post-reproductive age class;
                    d) a small reproductive age class and a large
                       pre-reproductive age class.

10.29     When studying the age structure of plant populations
          scientists use _____ of the plant to find
          age.
                    a) height;
                    b) weight;
                    c) biomass;
                    **d) diameter.**

10.30     Which of the following may influence a species' range?
                    a) climatic changes;
                    b) predation;
                    c) competition;
                    d) a and b, but not c.
                    **e) a, b, and c.**

**TRUE/FALSE**

10.31  **T**  F    It is difficult to apply the principles of
                   population ecology to plants, because a single
                   plant is not one unit, but is  a collection
                   of sub-populations.

10.32  **T**  F    The number of rats living in a city block would be
                   a measure of the population's crude density.

10.33  T  **F**    Territoriality in animals will result in a clumped
                   pattern of distribution.

10.34  T  **F**    The boundaries of a species' range are usually fixed and do not change.

10.35  **T**  F    An analysis of the effects of habitat change, competition, and predation in the distribution of a species is the subject matter examined in biogeography.

10.36  **T**  F    Organisms are distributed in space, but also in time.

10.37  T  **F**    The age structure of a population has little or no effect on the birth rates and death rates of the population.

10.38  T  **F**    Populations can be divided into two ecological periods, reproductive and post reproductive.

10.39  T  **F**    Among species that complete their life cycle in one year the pre-reproductive period has a significant effect on the rate of the population's growth.

10.40  **T**  F    In theory all continuously breeding populations should tend toward a stable age distribution.

10.41  T  **F**    Most natural populations reach a stable age distribution after a few years.

10.42  **T**  F    In growing populations the base of the age pyramid will be wide because the pre-reproductive segment of the population is large.

10.43  **T**  F    Population growth in wild animals may be influenced by environmental factors such as the availability of food and habitat.

10.44  T  **F**    It is possible to create good age structure diagrams for plant populations because it is easy to find the age of individual plants in a population by counting annual growth rings.

## MATCHING

10.45  __**A**__    Least significant period for annual plants.

10.46  __**A**__    Called the "Juvenile" period when examining plant populations.

10.47  __**C**__    The smallest period represented in the age structure diagram of India.

10.48 __C__ The largest period exhibited by the age structure of a declining population.

10.49 __B__ The period exhibited by the age structure that will be responsible for the number of young for future generations.

    A. Pre-reproductive.
    B. Reproductive.
    C. Post-reproductive.


## DISCUSSION

10.50 What type of factors may result in a population showing a clumped pattern of distribution? A uniform pattern of distribution?

10.51 What social and economic changes could you predict for the United States as the number of individuals in the older age classes increases?

10.52 Why is it that age pyramids and the predictions that can be drawn from them are more applicable for animals, especially birds and mammals, than for plants?

10.53 What is the difference between crude density and ecological density? Why is it important to have some measure of ecological density? Why is ecological density hard to determine?

10.54 Why is it important to examine both density and distribution when studying a population? How would you go about sampling the population size of a population of lupine growing in a meadow?

10.55 Three characteristics of all populations are density, distribution and age structure. Define each of these characteristics. Describe how these three attributes can be used to describe a population.

# Chapter 11

## Mortality, Natality, and Survivorship

### SENTENCE COMPLETION

11.1    **Demography** is the study of statistics that affect population growth.

11.2    **Mortality** is expressed as the probability of dying.

11.3    The growth or decline of a population is determined by the difference between the **natality** and the **mortality**.

11.4    The formula, **q=dt/Nt** is used to find the mortality rate.

11.5    The average number of years to be lived in the future by members of a population is known as **life expectancy**.

11.6    To obtain a clear and systematic picture of mortality and survival, demographers construct **life tables**.

11.7    **Time-specific** life tables record the mortality of each age class over a one year period.

11.8    **Mortality** curves plot mortality rates against age.

11.9    Mortality curves consist of two parts, the **juvenile** phase and the **post-juvenile** phase.

11.10   The mortality curve for mammals is called a **J-shaped** curve.

11.11   **Survivorship** curves plot logarithmic number of survivors against time.

11.12   **Type III** survivorship curves are typical of populations in which the mortality rate of juveniles is very high.

11.13   **Crude birthrate** is expressed as the births per 1000 per unit if time.

11.14   Age specific schedule of births is determined by dividing the **reproductively active females** into age classes and tabulating the number of **births** for each class.

11.15   **Fecundity** increases with size and age.

## MULTIPLE CHOICE

11.16    Mortality begins:
          **a) at fertilization;**
          b) after reproduction has taken place;
          c) shortly before old age;
          d) when individuals enter the post-reproductive cohort.

11.17    In mortality studies you begin the construction of a life table with a cohort of _____ individuals:
          a) 10;
          b) 100;
          **c) 1000;**
          d) 1,000,000.

11.18    In a life table $L_x$ represents:
          a) the number of time units left for all individuals to live for age x;
          **b) the average years lived by all individuals in each age category;**
          c) the fraction of the cohort that dies during age interval x;
          d) the number of individuals that died over a particular period.

11.19    Human life tables are widely used by:
          a) doctors;
          b) ecologists;
          **c) insurance companies;**
          d) geographers.

11.20    When taking a census of the age distribution of a population of animals, which of the following techniques would not be used?
          a) banding;
          b) capture and tagging;
          **c) measuring the volume of individuals;**
          d) aging individuals killed by hunters.

11.21    The life table approach in the study of plant populations is  useful when examining all except which of the following?
          a) the mortality and survival of seeds;
          b) the life cycle of annual plants;
          c) the population dynamics of marked seedlings of perennial plants;
          **d) the natality of annual populations;**

11.22    Yield tables for plants include all the following
         measurements except:
             a) age classes;
             b) diameter;
             c) basal area;
             d) volume;
             **e) there are several correct answers.**

11.23    Mortality curves include two parts:
             a) the reproductive and the post-reproductive
                phase;
             b) the reproductive and post-juvenile phase;
             c) the pre-juvenile and post-reproductive phase;
             **d) the pre-juvenile and juvenile phase.**

11.24    A mortality curve for a mammal population is roughly
         shaped like:
             a) a pyramid;
             b) the letter "S";
             c) a hexagon;
             **d) the letter "J".**

11.25    Type III survivorship curves would most likely be
         associated with:
             **a) many fish populations;**
             b) rodent populations;
             c) human populations;
             d) most mammal populations.

11.26    A survivorship curve that is strongly convex is typical
         of:
             **a) humans;**
             b) rats;
             c) sea urchins;
             d) perennial plants.

11.27    If mortality is constant at all ages the survivorship
         curve is:
             a) Type I;
             **b) Type II;**
             c) Type III;
             d) Type IV.

11.28    The greatest influence on the growth of a population
         is:
             a) death rate;
             b) mortality;
             **c) natality;**
             d) senility.

11.29    An animal population has 20 births per 10000 per year.
         Such a figure is a measurement of:
              **a) crude birthrate;**
              b) gross reproductive rate;
              c) net reproductive rate;
              d) age-specific schedule of births.

11.30    Data used to determine the aging pattern of animals
         collected during the hunting season will be biased in
         favor of _____ age classes.
              **a) older;**
              b) younger;
              c) pre-reproductive;
              d) rapidly reproducing age classes.


**TRUE/FALSE**

11.31  T  **F**    The growth or decline of a population is measured
                   by the difference between the number of births and
                   the number of immigrations.

11.32  T  **F**    Mortality begins when organisms reach a
                   reproductive age.

11.33  T  **F**    The mortality rate of a population is determined
                   by dividing the number of individuals that died
                   during a period of time by the number of new
                   births during the same period of time.

11.34  **T**  F    The age-specific mortality rate (q) is determined
                   by dividing the number of individuals that died
                   during a time (x) by the number of organisms alive
                   at the beginning of the time x.

11.35  **T**  F    A dynamic life table would provide you with data
                   concerning the fate of a group of organisms all
                   born at the same time.

11.36  T  **F**    Dynamic-life tables and time-specific life tables
                   are especially useful when studying the mortality
                   of insects because insects often have one breeding
                   season and all individuals will belong to the same
                   age class.

11.37  **T**  F    You cannot calculate life expectancy for an annual
                   species.

11.38  T  **F**    The bulk of a plant population's biomass is found
                   in its seeds.

171

11.39  **T**  F     Plant demographers have two levels of mortality to consider, that of a genet and of a metapopulation.

11.40  T  **F**     The juvenile phase of a mortality curve shows a very low rate of mortality.

11.41  T  **F**     If mortality rates are constant for all age groups in a population the survivorship curve will be concave.

11.42  **T**  F     A mortality curve for a mammal population will take the shape of the letter "J".

11.43  **T**  F     Natality is the main reason populations increase in size.

11.44  **T**  F     Germination in plants is equivalent to birth in animals.

11.45  T  **F**     Biologists agree that asexually produced plants represent natality.

## MATCHING

Answer the following questions concerning the survivorship of populations.

> A.  Type I
> B.  Type II
> C.  Type III

11.46  __A__  The type of pattern seen for human populations.

11.47  __C__  The type of pattern seen for most invertebrate populations.

11.48  __A__  Most of the offspring will survive to reach old age.

11.49  __B__   Individuals in this population have an equal chance of dying at any age.

11.50__B__   This is the type of pattern exhibited by most perennial plants.

## DISCUSSION

11.51      What is a life table?  What information is needed to assemble such a table?  What type of predictions can you make about populations from the information found in life tables?

11.52    Is mortality better defined as the probability of
         dying or the probability of surviving?  Which
         definition is of greater value to a demographer?

11.53    Distinguish the difference between the three types
         of survivorship curves.  What reasons can be given
         for the appearance of these different curves?

11.54    Plant populations present some special challenges
         when the dynamics of their mortality, natality, and
         survivorship are analyzed.  What information would
         you need to develop a plant life table?  How would
         you define mortality in plants?  How would you create
         a plant mortality curve?  How would you define
         natality in plants?

11.55    What is net reproductive rate?  How does it differ
         from the gross reproductive rate?  Which rate is of
         greater value when studying natality?

# Chapter 12

## Population Growth

**SENTENCE COMPLETION**

12.1    **Exponential growth** occurs when populations are not crowded.

12.2    The two major factors influencing population growth are **mortality** and **natality**.

12.3    If the number of births **equals** the number of deaths, a population experiences zero population growth.

12.4    All vertebrates have **overlapping generations** because parents continue to produce offspring even while some of their young begin to reproduce.

12.5    Population growth is influenced by the life history features of the population and by its **heredity**.

12.6    When an animal population is introduced into a new uncrowded environment they exhibit a population growth curve that is **J-shaped**.

12.7    The time required to double a population size is its **doubling time**.

12.8    Population growth is **density-dependent**, whereas exponential growth is **independent** of population density.

12.9    The **Allee effect** suggests that population growth is highest when a population is experiencing moderate densities.

12.10   The **carrying capacity** of a population is determined by its limiting resources.

12.11   Most populations become extinct because of changes in their **habitat**.

12.12   During the late **Permian** 90% of the shallow-water invertebrates became extinct.

12.13   Competition and resource depletion can act as **negative feedback** to populations and slow growth.

12.14   **Logistic** growth occurs when there is some limit placed on population growth by the environment.

174

12.15    Most human populations worldwide are experiencing
         **exponential** growth.

**MULTIPLE CHOICE**

12.16    An exponential growth curve is characteristic of:
           a) human populations;
           b) bacteria populations;
           c) fungal populations;
           **d) vertebrate populations introduced into a new
           environment.**

12.17    Which of the following factors is important when
         deciding carrying capacity?
           a) disease;
           b) starvation;
           c) competition;
           d) emigration;
           **e) all are important when deciding carrying
           capacity.**

12.18    Which statement is true?
           **a) most Western nations are experiencing zero
           population growth;**
           b) most Third World countries are experiencing
           zero population growth;
           c) environments can support sustained exponential
           growth;
           d) the exponential growth of a population is often
           followed by extinction of the population.

12.19    A bacterium can divide every half hour.  If we began
         with a single bacterial cell, how many bacteria would
         be in the population at the end of four hours if the
         population was growing exponentially?
           **a) 256;**
           b) 350;
           c) 5,280;
           d) 12.

12.20    If you begin with a single bacterium that divides every
         half hour, how many bacteria will there be at the end
         of four hours if there is a logistic population growth
         and the carrying capacity is 256?
           a) 350;
           b) 5,280
           c) 12;
           **d) 256.**

12.21    Which of the following animals does not exhibit
         oscillations in their population size?
                 a) lynx;
                 b) lemmings;
                 c) snowshoe hare;
                 **d) reindeer.**

12.22    Which of the following characteristic of a species
         would suggest a high rate of extinction?
                 a) small body size;
                 b) large geographic range;
                 c) high genetic plasticity;
                 **d) large body size.**

12.23    The most important cause of species extinction today
         is:
                 **a) habitat destruction;**
                 b) climatic change;
                 c) changes in the oceanic currents;
                 d) changes in predator pressure.

12.24    In exponential growth there is_____ feedback
         which means that the larger the population becomes the
         faster it grows.
                 a) negative;
                 **b) positive.**

12.25    The two major factors influencing population growth
         are:
                 a) growth potential and exponential growth;
                 **b) mortality and natality;**
                 c) survivorship and emigration;
                 d) emigration and immigration.

**TRUE/FALSE**

12.26  **T**  F    If the births within a population are greater than
the               deaths, the population increases.

12.27  T  **F**    An individual that leaves a population is called
                   an immigrant.

12.28  **T**  F    Disease is a density dependent factor.

12.29  T  **F**    When a species is introduced into a new habitat
                   the species experiences growth without
                   environmental resistances.

12.30  **T**  F    The carrying capacity of an ecosystem is
                   determined by the availability of space, water,
                   light and nutrients.

176

12.31  **T**  F     The maximum population size that is supported by a particular environment is called the carrying capacity.

12.32  T  **F**     Growth of populations results from negative feedback.

12.33  T  **F**     The lynx and the snowshoe hare show oscillation intervals of approximately three to four years.

12.35  **T**  F     Habitat destruction is the major cause of species extinction.

12.36  **T**  F     Competition for environmental resources may act as a negative feedback and slow population growth.

12.37  **T**  F     Sparse populations often face increased death rates because predation pressure on the remaining individuals is high.

12.38  T  **F**     Population extinction has been more-or-less spread evenly over Earth's geologic history.

12.39  **T**  F     No environment can support sustained exponential growth.

12.40  **T**  F     Because environments are constantly changing and because resources are finite in availability population growth declines with increasing population density.

12.41  **T**  F     Weather is a density-independent factor that limits population growth.

## MATCHING

**REFER TO THE TABLE BELOW**

12.42 __A__     At this point on the graph, the rate of the population increase is slow.

12.43 __B__     The rate of increase is accelerating.

12.44 __D__     The carrying capacity has been reached.

12.45 __C__     The inflection point is reached.


## DISCUSSION

12.46     Some populations have very high birth rates.  What regulates the growth of populations?

12.47     What is carrying capacity?  What factors would be involved in deciding carrying capacity?

12.48     What role does extinction play in the history of populations?  Extinction rate has accelerated in recent years.  Why?

12.49     What is the difference between exponential growth and logistic population growth?

12.50     What is meant by net reproductive rate, finite rate of increase, and annual rate of increase?

# Chapter 13

Population Regulation

## SENTENCE COMPLETION

13.1    **Intraspecific competition** refers to the competition
        among individuals of the same species.

13.2    **Scramble** competition occurs when all the individuals
        within a population share a resource equally.

13.3    If one of two individuals within a population
        establishs a claim to an environmental resource **contest**
        competition will result.

13.4    Some plants respond to density by reducing their growth
        rate.  This phenomena is called **phenotypic plasticity.**

13.5    Crowded conditions increase the social contact within a
        population and lead to **stress.**

13.6    Hormonal changes under stress may weaken the **immune**
        system making individuals susceptible to disease.

13.7    Chemical substances released by an animal into the
        environment that can influence the behavior of others
        in the population are called **pheromones.**

13.8    Many animals avoid stress by seeking less crowded areas
        of their habitat.  Such a movement away from a crowded
        area is called **dispersal.**

13.9    **Social hierarchy** refers to the social organization of
        some animals based upon intraspecific aggression.

13.10   The male leader in a pack of wolves is called the **alpha**
        male.

13.11   **Territoriality** refers to a behavior in which an
        individual claims an area for itself and excludes
        others from sharing it.

13.12   The area that an animal moves through regularly but
        does not defend is its **home range.**

13.13   The home range of herbivores **increases** as body weight
        **increases.**

13.14   Some plants release **phenolic compounds** into the soil
        that retard the germination of seeds.

13.15    As the size of a territory **increases** the amount of
         energy expended in defense is **increased**.

**MULTIPLE CHOICE**

13.16    Which of the following is a population control that is
         density-dependent?
                a) weather;
                **b) disease;**
                c) earthquakes;
                d) forest fires.

13.17    Which statement does not apply to scramble competition?
                a) all individuals share resources equally;
                b) this is more likely the type of competition
                   seeds released by a plant must face;
                c) the population density remains below what the
                   environment can support;
                **d) this type of competition allows individuals to
                   tap efficiently  natural resources with little
                   or no waste.**

13.18    Which statement best describes contest competition?
                a) resources are shared equally by all members of
                   a population;
                b) most individuals in the population get less
                   than what they need from their environment;
                **c) a small proportion of the population gets
                   adequate resources and they produce the next
                   generation;**
                d) contest competition is best described as a
                   free-for-all with the resources as the prize.

13.19    As the density of a poikilotherm vertebrate _____,
         their rate of growth _____.
                **a) increases, decreases;**
                b) increases, increases;
                c) decreases, decreases;
                d) there is no correct answer.

13.20    As the density of a population in scramble competition
         _____, individuals _____ their food intake.
                a) increases, increase;
                **b) increases, decrease;**
                c) decreases; increase.

13.21    Phenotypic plasticity is most likely to be exhibited by
         _____ populations.
                a) vertebrate;
                b) insect;
                c) fish;
                **d) plant.**

13.22    As plant populations _____ in size the biomass in
         the remaining individuals _____ .
         a) decrease, decreases;
         b) increase, increases;
         **c) decrease, increases;**

13.23    Which statement about stress is incorrect?
         a) as population density increases, stress
            increases;
         b) stress triggers changes in the endocrine system
            of animals;
         **c) stress can stimulate growth and increase the
            rate of reproduction;**
         d) stress may result in profound changes in the
            immune system.

13.24    Stress can result in _____ births and _____ infant
         mortality.
         a) increased, increased;
         **b) decreased, increased;**
         c) increased, decreased;
         d) decreased, decreased.

13.25    Which statement is correct?
         a) when a resource shortage causes dispersal to
            take place usually the reproductive adults are
            forced out;
         **b) dispersal allows for some populations to locate
            new habitats;**
         c) dispersal functions as an important mechanism
            regulating the growth of populations;
         d) seeds scattered near the adult plant are more
            likely to survive than those that grow some
            distance from the parent.

13.26    Social hierarchies:
         a) are restricted to invertebrates and rarely
            appear in vertebrate populations;
         b) become disruptive to normal social interaction;
         **c) stabilize relationships and minimize fighting;**
         d) are not believed to play a role in population
            regulation.

13.27    The alpha male and the alpha female:
         a) feed first;
         b) are the reproducing pair;
         c) are subordinate to the omega male and omega
            female;
         **d) a and b, but not c;**
         e) a, b, and c.

13.28    Chemical secretions that are released by an animal into
         the environment that can alter the behavior of others
         of the populations are called:
                   a) hormones;
                   **b) pheromones;**
                   c) floaters;
                   d) phenolic compounds.

13.29    If food was a limited resource, you would expect the
         size of an animal's territory to:
                   a) decrease in size;
                   **b) increase in size;**
                   c) not change in size because territories are
                      density independent.

13.30    Which statement about home range is incorrect?
                   **a) the size of the home range always remains the
                      same;**
                   b) carnivores have a larger home range than
                      herbivores;
                   c) males have larger home ranges than females;
                   d) the home range of an omnivore increases as the
                      biomass of the individual increases.

## TRUE/FALSE

13.31  **T**  F   Density-dependent population regulations influence
                  a population in proportion to its size.

13.32  T  **F**   Competition for resources is a density-independent
                  influence.

13.33  **T**  F   Contest competition results in some individuals in
                  the population getting more resources than others.

13.34  **T**  F   Scramble competition results in the wasting of
                  environmental resources.

13.35 **T**  F   As food quality decreases scramble competition among
                 grazers becomes more intense.

13.36  **T**  F   Increasing the number of aggressive encounters
                  within a population will trigger profound changes in
                  the hormone production of some individuals.

13.37  T  **F**   Stress may act as a regulatory mechanism on plant
                  populations.

13.38  **T**  F   Individuals that disperse even when population
                  density is low are genetically predisposed to
                  disperse.

13.39  T  **F**  The probability of seed survival decreases with the distance from the parent plant.

13.40  **T**  F  Social hierarchies create a harmonious social organization and reduce the level of aggression in the group.

13.41  **T**  F  Territorial behavior includes some form of advertisement, threat, and combat.

13.42  **T**  F  As the size of an animal's territory decreases the amount of energy it expends to defend it increases.

13.43  **T**  F  Territories are defended, but home ranges are not.

13.44  T  **F**  Home ranges have rigid boundaries and do not change in size.

13.45  **T**  F  If you place too many fish in an aquarium, they will grow slowly and will never reproduce.

**MATCHING**

A.  density-dependent regulation
B.  density-independent regulation

13.46  __**B**__  heavy winter freeze

13.47  __**A**__  predation

13.48  __**B**__  pesticides

13.49  __**A**__  smallpox introduction by European into the natives of Hawaii

13.50  __**A**__  parasitism

**DISCUSSION**

13.51  Describe some density-independent factors that would influence the size of the mosquito population in a lake.  Would any of these factors operate to limit the size of human populations?

13.52  What is the evolutionary value of social hierarchy and territoriality?  Do human populations exhibit these two behaviors?

13.53  What is competition?  How are scramble competition and contest competition alike?  How are they different?

13.54    How does stress and dispersal regulate population
         growth?

13.55    What types of population regulating mechanisms are
         operating on plants?  Are these mechanisms unique to
         plants or can they also regulate animal populations?

# Chapter 14

## Life History Patterns

14.1    The selective processes by which organisms achieve fitness are called **life-history strategies.**

14.2    Success or failure of a population is measured by the number of successful **offspring.**

14.3    The amount of time and energy that an organism puts into reproduction is called the **reproductive effort.**

14.4    **Semelparity** is as type of reproduction in which an organism expends all its energy in growth and reproduction and then dies.

14.5    If an animal produces many offspring it will provide **no** parental care.

14.6    Robin chicks are helpless at hatching and must be fed by their parents.  They are **altricial chicks.**

14.7    Grazing animals have young that can forage for themselves shortly after birth.  They are **precocial.**

14.8    Birds of the same family have **smaller** clutches at low latitudes than at **high** latitudes.

14.9    Among poikilotherms, **fecundity** increases with the size and age of an organism.

14.10   The **r**-strategy for reproduction is characteristic of populations living in harsh, unpredictable environments.

14.11   **K**-strategists are limited by the resources of their environment and so they remain close to their carrying capacity.

14.12   Holly is a **dioecius** plant.  There is a separate male and female plant.

14.13   **Hermaphrodites** have both male and female sex organs in one individual.

14.14   **Monoecy,** a condition limited to plants occurs when a single plant contains both male and female flowers.

14.15    **Monogamy** involves a pair bond between one male and one
         female. **Polygamy** occurs when an individual has two or
         more mates.

## MULTIPLE CHOICE

14.16    Which of the following statements is false?
         a)    the more energy an organism uses in
               reproduction the less it has for growth;
         b)    non-reproducing females devote most of their
               energy to growth;
         c)    most insects invest all their energy into
               growth and development and into one massive
               reproductive effort;
         d)    **timing of reproduction to occur early in life
               will increase survivorship but it reduces the
               potential for future reproduction.**

14.17    As an individual _____ its reproductive effort it
         _____ its survivorship.
         a)    **increases, reduces;**
         b)    increases, increases;
         c)    increases, does not change;
         d)    decreases, reduces.

14.18    Parents that produce many young:
         a)    reproduce often;
         b)    exhibit intensive parental care;
         c)    **invest little energy in parental care;**
         d)    usually inhabit stable, well-established
               environments.

14.19    Plants found in small environments produce:
         a)    **large seeds;**
         b)    many seeds;
         c)    small seeds;
         d)    seeds with little stored energy.

14.20    A population in which the females produce one large
         clutch early in their life is:
         a)    **r-selected;**
         b)    K-selected;
         c)    iteroparous;
         d)    altricial.

14.21    Which of the following is a hermaphrodite?
         a)    **earthworm;**
         b)    trout;
         c)    lizard;
         d)    jack-in-the-pulpit.

14.22    When one male mates and stays with one female, the
         relationship is:
         a)   polygamous;
         **b)   monogamous;**
         c)   monoecious;
         d)   promiscuous.

14.23    Sexual reproduction:
         a)   allows for new genetic combinations;
         b)   increases genetic variability in the species;
         c)   prevents the accumulation of harmful
              mutations;
         d)   a and b, but not c;
         **e)   a, b, and c.**

14.24    Dioecious species:
         a)   are monoeious;
         b)   are hermaphrodites;
         c)   possess the sex organs of both male and
              female;
         **d)   include plants where there is a separate male
              and female individual.**

14.25    Many young is characteristic of:
         a)   long-lived mammals;
         b)   perennial plants;
         c)   K-selected populations;
         **d)   annual plants.**

**TRUE/FALSE**

14.26  T  **F**   When ecologists refer to "strategies" they are
               suggesting that plants and animals can make plans
               toward the achieving of a specific goal.

14.27  **T**  F   The life-history strategy is the outcome of
               evolution and represents a pattern developed
               through natural selection.

14.28  **T**  F   The more energy an organism spends on
               reproduction, the less it has for homeostasis and
               growth.

14.29  **T**  F   Iteroparous organisms, which reproduce later in
               life, have a high survivorship but a reduced
               fecundity.

14.30  T  **F**   The human baby is precocial because it is born
               helpless and requires intensive parental care.

14.31  **T**  F   Short-lived annual plants produce many seeds.

14.32  T  **F**  Oak trees produce a few large seeds containing large amounts of stored energy.  They reflect the characteristics of plants adapted to harsh environments.

14.33  T  **F**  Natural selection favors slow reproduction and small clutch and litter size in animals living in temperate regions.

14.34  T  **F**  Among all homotherms fecundity increases with size and age.

14.35  **T**  F  Sexual reproduction is important to organisms because it allows mixing of the gene pool and increase genetic variability.

14.36  T  **F**  Earthworms are monoecius.

14.37  T  **F**  Hermaphrodites maintain genetic variability by self-fertilization.

14.38  **T**  F  To maintain fitness males should mate with as many females as possible.

14.39  **T**  F  The evolution of dioecy from male-dominated hermaphroditic flowers is evidence of sexual selection among plants.

14.40  **T**  F  Mammals with a high metabolic rate have a higher fecundity than mammals with a low metabolism.

## MATCHING

A.  r-selected species
B.  K-selected species

14.41  __B__  slow maturation

14.42  __B__  reproduce often

14.43  __A__  rapid maturation

14.44  __A__  characteristic of colonizers of new environments

14.45  __A__  oysters produce two million young at once

## DISCUSSION

14.46  Polygyny is much more prevalent in mammals than in birds.  Why is this the case?  Why are carnivorous mammals usually monogamous?

14.47    What selective pressures might explain the evolution of polyandry?

14.48    Is there an advantage for hermaphrodism over separation of the sexes?  Why do most multicellular animals reproduce sexually?

14.49    Many animals conduct elaborate courtships preceding copulation.  What is the adaptive significance of courtship behavior?

14.50    Most animals reproduce during specific seasons, rather than all year.  What adaptive advantage is there for seasonal reproduction?

# Chapter 15

## Interspecific Competition

### SENTENCE COMPLETION

15.1    Interactions between members of different species are called **interspecific** interactions.

15.2    In **commensalism**, one population receives benefit from interacting, while the other population receives no benefit, but is not harmed while interacting.

15.3    **Amensalism** involves one population being harmed by an interaction and the other neither being harmed or receiving any benefit.

15.4    In **obligatory mutualism** both populations benefit by interacting and this relationship is essential for the survival of both.

15.5    **Predation** refers to the killing and consumption of prey.

15.6    **Parasitism** is not beneficial to the host, but it is of survival importance to the parasite.

15.7    **Competition** offers no benefit to either interacting population.

15.8    Certain wasps lay their eggs in the body of an insect. When the eggs hatch the young wasps eat the insect. This combination of predation and parasitism is called **parasitoidism.**

15.9    When two or more species are seeking out the same environmental resource **interspecific** competition takes place.

15.10   Because two competing species have to share a limited resource the **carrying capacity** of each species **decreases.**

15.11   If competing species share a resource, **coexistence** results, even if the competition reduces the fitness of both populations.

15.12   The production of toxins which when released into soil inhibit the growth of another population is an example of **allelopathy.**

190

15.13    The term **niche** is used to describe all the
         environmental requirements of a population.

15.14    A **fundamental niche** is the set of environmental
         conditions that a population is capable of using
         limited by its genetic makeup.

15.15    The **realized niche** is the set of environmental
         conditions that a population actually uses.

## MULTIPLE CHOICE

15.16    Symbiosis is an interspcific interaction between two
         species:
         **a)  which live together and come in contact with
              each other;**
         b)  which harm each other when they come in
             contact;
         c)  which must live together to be successful;
         d)  which neither harm or benefit each
             population.

15.17    If + is the symbol for a positive effect and - a
         detrimental effect, then parasitism would be symbolized
         by:
         a)   ++;
         **b)   + -;**
         c)   - -;
         d)   0 +.

15.18    Commensalism is an interaction between two species in
         which:
         a)  both benefit;
         b)  neither benefit;
         **c)  one benefits and the other receives no
              benefit or harm;**
         d)  both are harmed.

15.19    Obligatory mutualism occurs when:
         a)  two populations interact with each other;
         b)  one population must interact with the other,
             but one population is harmed by the
             interaction;
         **c)  interaction is essential for the success of
              both populations;**
         d)  one population survives but the other
             population's fitness is reduced.

15.20    Parasitism and _____ involve an interaction in
         which the fitness of one population is reduced and the
         fitness of the other is increased.
         a)    competition;
         b)    commensalism;
         **c)    predation;**
         d)    mutualism.

15.21    G. F. Gause examined competition in:
         a)    ants;
         b)    squirrels;
         c)    parasites;
         **d)    protozoans.**

15.22    The competitive exclusion principle states:
         a)    two populations can coexist as long as they
               occupy separate niches;
         **b)    two populations that are competitors cannot
               coexist;**
         c)    two populations that compete will genetically
               change over a period of time;
         d)    two populations that compete will increase
               their carrying capacity.

15.23    When two or more organisms use a portion of the same
         resource simultaneously _____ occurs.
         **a)    niche overlap;**
         b)    predation;
         c)    commensalism;
         d)    niche width.

15.24    The niche of a population is the:
         a)    environment in which it lives;
         b)    total of all the interactions it has with
               other populations;
         c)    the range in which it roams;
         **d)    the sum of all its environmental
               requirements.**

15.25    Allelopathy is a form of competition in:
         a)    mammals;
         b)    birds;
         c)    protozoa;
         **d)    plants.**

**TRUE/FALSE**

15.26  T  **F**   Parasitoidism is a combination of two types of
                  interactions, parasitism and competition.

15.27  **T**  F   Flowering plants are pollinated by insects.  This
                  is an example of mutualism.

192

15.28  **T**  F    Crows follow a farmer as he plows a field because they can eat insects that he uncovers.

15.29  T  **F**    Intraspecific competition favors specialization, whereas interspecific competition favors generalization.

15.30  **T**  F    Two complete competitors cannot coexist, one will become extinct.

15.31  **T**  F    The best examples of unstable equilibrium occur in plants.

15.32  T  **F**    Allelopathy describes a process by which rodents release a plant growth inhibitor into the soil when they urinate.

15.33  **T**  F    Two rodents eat the same plant, but each rodent eats a different part of the plant.  Such an observation suggests a resource partitioning is taking place.

15.34  T  **F**    The habitat of a population when combined with its niche comprise the hypervolume.

15.35  **T**  F    An individual free from competition or interference from other organisms could occupy its fundamental niche.

15.36  T  **F**    Resource partitioning is likely the outcome of intraspecific competition.

15.37  T  **F**    Ecologists usually examine as many niche dimensions as possible when trying to describe the realized niche of a species.

15.38  **T**  F    It is unlikely that two species will have exactly the same niche requirements.

15.39  T  **F**    Niche compression is likely to take place when a competing species becomes extinct and the species remaining moves into habitats once occupied by the extinct species.

15.40  **T**  F    Organisms may undergo niche shift and thereby reduce competition.

## MATCHING

A. amensalism
B. mutualism
C. commensalism
D. parasitism
E. competition

15.41 __E__  Two trees grow side by side, each reaching for light.

15.42 __A__  Penicillium produces a toxin that will kill some bacteria.

15.43 __B__  A termite cannot digest wood without the aid of a protozoan that lives in its intestine.

15.44 __C__  A crab attaches a sea anemone to its shell to camouflage its body. The sea anemone gets no benefit from the interaction, but it is not harmed by the crab.

15.45 __D__  A tapeworm lives in the intestine of a dog.

15.46 __E__  Caribou and reindeer feed on the same tundra plants.

15.47 __B__  A small bird eats the ticks off an iguana's back.

15.48 __A__  An elephant steps on an ant.

15.49 __D__  A flea sucks the blood from a cat.

15.50 __C__  A woodpecker stores acorns by depositing them in small holes it drills in the bark of a tree.

## DISCUSSION

15.51  What is a niche? What is the difference between fundamental niche and realized niche?

15.52  What are the similarities and what are the differences between predation and parasitism? Why don't parasites kill their host?

15.53  What are four potential outcomes of competition as described by Alfred Lotka and Vittora Volterra? Describe two experiments that were carried out in laboratories to investigate competition.

15.54  Is it possible for potentially competing species to coexist?

15.55    What is the competitive exclusion principle?  Why is it
         difficult to demonstrate competition under natural
         conditions?

## SENTENCE COMPLETION

16.1    When a deer browses on the leaves of a shrub the
        activity is called **herbivory.**

16.2    **Cannibalism** would occur when the predator and the prey
        belong to the same species.

16.3    In the Lotka-Volterra model of predation the assumption
        is made that the **prey** population grows **exponentially.**

16.4    A.H. Nicholson and W. Bailey proposed the idea that the
        abundance of prey influences the **reproductive rate** of
        the predator.

16.5    **Functional response** assumes that a predator will eat
        more prey as the prey population increases.

16.6    A **Type III** functional response can stabilize a prey
        population because the rate of predator attacks varies
        with the density of the prey.

16.7    New prey in a predator's habitat are not killed.  Such
        an observation is support for the **search image
        hypothesis.**

16.8    If the prey population for which a predator has a
        strong preference decreases the predator may turn to an
        alternate, more abundant prey.  The process is called
        **switching.**

16.9    A direct numerical response takes place when the number
        of predators **increase** as the density of prey
        increases.

16.10   Cannibalism is a type of **intraspecific** predation.

16.11   A grass is well adapted to withstand **grazing.**  The
        grazer will remove old plant tissue first, but the
        plant will grow because the **meristem** is located near
        the ground.

16.12   To build animal tissue a herbivore needs forage rich in
        **nitrogen.**

16.13   **Mimicry** results when a palatable prey species resembles
        a distastful species.

16.14    Predators have evolved three general patterns of
         hunting: **ambush, stalking, and pursuit.**

16.15    The means by which an animal secures its food is its
         **foraging strategy.**

## MULTIPLE CHOICE

16.16    The Lotka-Volterra model of predation suggests that:
         a)    when predator populations increase, prey
               populations decrease;
         b)    when predator populations decrease, prey
               populations increase;
         c)    when predator populations increase, prey
               populations increase;
         **d)    a and b, but not c;**
         e)    a,b, and c.

16.17    G. F. Gause's experiments on predation suggests that:
         a)    paramecia develop defenses against continual
               predation;
         b)    density of predators increases as the density
               of prey decreases;
         **c)    a predator may exterminate its prey and then
               die of starvation.**

16.18    Studies on predator-prey interactions suggest that:
         a)    predators can survive when prey populations
               are low;
         **b)    a self-sustaining predator-prey population
               requires an immigration of new prey to
               maintain itself;**
         c)    the intensity of predation is the single-most
               important factor in influencing the
               oscillation of predator-prey populations;
         d)    all of the statements are correct.

16.19    In a Type I functional response:
         **a)    the number of prey taken increases as prey
               density increases;**
         b)    the number of pre taken increases at a
               decreasing rate to a maximum value;
         c)    the number of prey taken is low at first then
               increases until it approaches an upper limit;
         d)    handling time is a major component.

16.20    An inverse numerical response to predation has taken
         place when:
         a)    the number of predators increases as the
               number of prey increases;
         b)    the predator population does not change as
               the number of prey increases;
         **c)    predators increase sharply for a short time
               then decline rapidly as the prey density
               increases.**

16.21    How can a herbivore affect the fitness of a plant?
         a)    they may increase the plant's biomass by
               eating plant tissue;
         b)    they may cause a growth of new foliage that
               will add to the nutrient reserves of the
               plant;
         c)    they may cause plant buds to go dormant;
         **d)    they may alter the growth form of the plant.**

16.22    The distasteful pipevine swallowtail and the tasty
         black swallowtail are similar in appearance.  This
         observation is an example of:
         a)    mobbing;
         b)    distraction display;
         **c)    mimicry;**
         d)    cryptic coloration.

16.23    Which of the following methods of hunting has a low
         frequency of success but requires minimal energy?
         **a)    ambush;**
         b)    stalking;
         c)    pursuit.

16.24    Which of the following is an example of cryptic
         coloration?
         **a)    the spots on the coat of a fawn;**
         b)    the distinctive yellow and black pattern of a
               wasp;
         c)    the color pattern of a viceroy butterfly;
         d)    the armor coat of a beetle.

16.25    Which of the following characteristics may serve to
         protect a plant from a predator?
         a)    hair leaves;
         b)    spines;
         c)    cyanogenic compounds;
         d)    a and c, but not b;
         **e)    a, b, and c.**

16.26   **T**   F   Lotka and Volterra concluded that the growth rate of a predator was influenced by the density of a prey population.

16.27   **T**   F   Gause showed that predators may overexploit their prey, exterminate their food, and then die.

16.28   T   **F**   A self-sustaining predator-prey population can be maintained without the immigration of prey.

16.29   T   **F**   The intensity of predation is the most important factor deciding the ability of a prey population to recover from the  pressures of predation.

16.30   **T**   F   The idea of functional response is based on the assumption that a predator will take more prey as the population density of prey increases.

16.31   **T**   F   In Type I functional response the number of prey taken per predator increases as prey density increases.

16.32   T   **F**   Type II functional response often acts as a force that will stabilize a prey population.

16.33   **T**   F   Compensatory predation occurs when prey numbers increase above the threshold of security and the surplus animals become more susceptible to predation through intraspecific competition.

16.34   T   **F**   When a new species is introduced into a predator's habitat it will become an accepted prey in a short time.

16.35   **T**   F   Aldo Leopold said that alternate prey species may act as a buffer reducing the predation of a game species.

16.36   **T**   F   When a predator selects an alternate, more abundant prey over its usual prey switching has taken place.

16.37   T   **F**   Ecologists believe that cannibalism is very rare in animal populations since it occurs when normal food supply is inadequate to support the population's numbers.

16.38   **T**   F   The two components of foraging strategy are optimal diet and optimal foraging efficiency.

16.39  **T**  F   When herbivores feed on young tender plant tissue they are removing large amounts of nutrients form the plant and are lowering the plant's fitness.

16.40  **T**  F   Annuals may produce chemical substances that interfere with the metabolism of herbivores.

## MATCHING

A. Type I Functional Response
B. Type II Functional Response
C. Type III Functional Response

16.41  __C__   Also called compensatory predation.

16.42  __B__   Handling time decides the maximum number of prey that can be taken by a predator.

16.43  __B__   The number of prey taken by the predator rises as prey density increases up to a maximum value.

16.44  __A__   Predator eats the prey until the predator is full.

16.45  __A__   A ladybug eats aphids on a rose bush until it is satiated.

## DISCUSSION

16.46   Lotka, Volterra and others developed mathematical models to express the relationship between predator and prey. What are the major hypotheses presented by these models? Your textbook points out that these models are simplistic. What were the reasons for such a statement?

16.47   Coral snakes are venomous. Their colorful markings look similar to the markings of a few non-venomous species. Under what conditions could these similarities be of value? How did they evolve?

16.48   What is the difference between ambush, pursuit, and stalking? Describe some conditions that may explain how each of these mechanisms evolved.

16.49   Plants are at the mercy of the herbivores. Would you agree or disagree with this statement? Are there plants defenses that could protect them from foragers?

16.50   Vegetation, herbivores, and predators are involved in a complex set of relationships. Describe a three-level interaction that has been studied to prove this point.

# Chapter 17

## PARASITISM AND MUTUALISM

### SENTENCE COMPLETION

17.1    **Parasitism** is an interspecific interaction in which two organisms live together but one two gets its nourishment  from the tissues of the other.

17.2    Viruses and bacteria belong to a subcategory of parasites called **microparasites**.  Fleas and ticks are in another subcategory called **macroparasites**.

17.3    Organisms that carry or transmit a parasite from one host to another are called **vectors**.

17.4    The **definitive host** is the one in which the parasite becomes an adult.  The **intermediate** host is the one that harbors the parasite as it is developing.

17.5    Mistletoes are classified as **hemiparasites**.

17.6    The cowbird lays its eggs in the nest of another bird. This behavior is an example of **temporary obligatory parasitism.**

17.7    When two organisms interact physically and their interaction is obligatory for the survival of the two organisms, the association is called **mutualism**.

17.8    The dispersal of seeds by animals is a type of nonobligatory mutualism called **facultative mutualism**.

17.9    Often **frugivores** scatter seeds as they eat the fruits of some types of plants.

17.20   The virulence of a pathogenic organism is matched by the evolution of resistance in the host species. The process by which this event takes place is **coevolution.**

### MULTIPLE CHOICE

17.21   Parasites:
      a)    are always agents of disease;
      b)    are all microscopic in size;
      **c)    may be mutualistic and needed by their host;**
      d)    are always animals.

17.22    Macroparasites:
        a)    are small and microscopic in size;
        b)    are always ectoparasites;
        c)    have a short generation time;
        **d)    rarely multiply in their host.**

17.23    Microparasites:
        **a)    have a short generation time;**
        b)    include rusts and smuts;
        c)    are always ectoparasites;
        d)    are the vectors for many diseases.

17.24    Which of the following is an example of the direct transmission of a parasite?
        **a)    parasitic nematodes are transmitted from one host to another;**
        b)    white-tailed deer may ingest a snail in which a roundworm lives;
        c)    white pine blister rust must spend part of its life cycle on an intermediate host shrub;
        d)    all of the above are examples of direct transmission.

17.25    Which statement is true?
        a)    microparasites require a low host density to persist;
        b)    host immunity will increase the parasite numbers;
        **c)    parasites spread by indirect transmission may persist for a long time in low host population densities;**
        d)    rabies will confer long-term immunity on its host.

17.26    Plants may discourage the consumption of their unripe fruit by which of the following methods?
        a)    producing chemicals that will repel a frugivore;
        b)    having an unpalatable texture when eaten;
        c)    possessing a hard seed coat;
        d)    having a fruit color that will blend with the plant's leaves;
        **e)    all these are correct answers.**

17.27    Which of the following strategies is typical of plants
         living in temperate regions?
         a)    their fruits are very succulent and high in
               sugars;
         b)    their seeds are resistant to the digestive
               enzymes of frugivores;
         c)    they produce few seeds and those that are
               produced are very large;
         **d)    a and b, but not c;**
         e)    a, b, and c are correct.

17.28    Nectivores:
         a)    visit plants to pollinate them;
         b)    are specialists usually visiting only one
               species of plants;
         c)    are generally birds and small mammals;
         d)    all the answers are correct;
         **e)    none of the answers are correct.**

17.29    Nonsymbiotic mutualism may have evolved from:
         a)    parasite-host relationships;
         **b)    exploitation of one population by another;**
         c)    predator-prey relationships;
         d)    commensalism.

17.30    Which of the following is not an example of symbiotic
         mutualism?
         a)    mycorrhizae attached to the roots of a plant;
         b)    an algae and a fungus living together;
         c)    bacteria in the human small intestine
               producing vitamins;
         **d)    bees pollinating flowers.**

**TRUE/FALSE**

17.31  T  **F**    All parasites cause disease.

17.32  **T**  F    A flea is an ectoparasite.

17.33  T  **F**    Direct transmission from one host to another by
                   direct contact is the usual way macroparasites are
                   spread.

17.34  **T**  F    Lyme disease is caused by a microparasite.

17.35  **T**  F    Holoparasites lack chlorophyll and take water and
                   nutrients from the roots of their host plant.

17.36  **T**  F    The transmission of a parasite is dependent on the
                   density of the hosts.

17.37  T  **F**   Immunity of a host may eliminate a parasite
                   population.

17.38  **T**  F   To control the spread of rabbits in Australia a
                   viral infection was introduced into the
                   population.

17.39  T  **F**   Termites cannot live without a small protozoa
                   that resides in their digestive tract.  This
                   relationship is an example of facultative
                   mutualism.

17.40  **T**  F   Symbiotic mutualism might evolve from other
                   interspecific interactions such as parasite-host,
                   commensalism, and predator-prey.

## MATCHING

a. ectoparasite
b. endoparasite

17.41  __A__   tick

17.42  __B__   tapeworm

17.43  __B__   Elmeria

17.44  __B__   Lyme disease

17.45  __A__   dodder

## DISCUSSION

17.46   What is mutualism?  How does obligatory symbiotic
        mutualism differ from obligatory nonsymbiotic
        mutualism?  Give examples of each.

17.47   What is the difference between true parasitism and
        social parasitism?

17.48   How would you define coevoultion?  How might this
        idea apply to our understanding of parasite-host
        relationships?

17.49   How are parasites transmitted to new hosts?  What role
        does an intermediate host play in the life cycle of a
        parasite?

17.50   What role does a parasite play in the regulation of a
        host population?

# Chapter 18

## Human Control of Natural Populations

### SENTENCE COMPLETION

18.1    Any animal that is undesirable is a **pest**.

18.2    A **weed** is a plant that is growing somewhere it is not wanted.

18.3    Most unwanted animals possess **r** characteristics such as high rate of increase and small size.

18.4    After World War II, **organic insecticides**, chemicals containing carbon, were developed to control pests.

18.5    Non-selective insecticides, which kill a variety of insects are called **broad spectrum**.

18.6    **Herbicides** are organic pesticides used to control plants.

18.7    The gypsy moth has been controlled by baiting traps with **pheromones** that attract the males of the population.

18.8    **Integrated pest management** is a holistic approach to the control of pests.

18.9    **Sustained yield** is the highest rate at which a population can be harvested without impairing its ability to be renewed.

18.10   If the **maximum** sustained yield is exceeded the population being harvested will decline.

18.11   **Optimal** sustained yield considers biological and sociological aspects when determining the number of individuals that can be harvested from a population.

18.12   **Habitat** destruction is the major cause of the decline of animal and plant species.

18.13   **Cropping** may be required to make sure that overpopulation of a threatened species does not exceed the carrying capacity of its habitat.

18.14   To guarantee the success of a wildlife management program protection of the habitat must be accompanied by **education** so as to gain public support.

18.15    It may be possible to breed some plants for **genetic resistance** and by that successfully inhibit insect populations.

<u>**MULTIPLE CHOICE**</u>

18.16    Which of the following could be classified as a pest?
         a)    insects;
         b)    birds;
         c)    mammals;
         d)    frogs;
         **e)    all of the above.**

18.17    Pyrethrin is extracted from:
         **a)    chrysanthemums;**
         b)    tomatoes;
         c)    orchids;
         d)    ragweed.

18.18    Malathion and Parathion are examples of:
         a)    chlorinated hydrocarbons;
         **b)    organophosphates**
         c)    carbamates;
         d)    herbicides.

18.19    _____ would persist in the environment longer than organophosphates.
         a)    pyrethrins;
         b)    carbamates;
         **c)    chlorinated hydrocarbons;**
         d)    diflubenezuron.

18.20    Which of the following is an example of a herbicide?
         a)    2,4-D;
         b)    2,4,5-T;
         c)    alrazine;
         **d)    all of the above.**

18.21    Which of the following was an important factor encouraging the development of most insecticides?
         a)    their toxicity;
         b)    the ease of application;
         c)    their low cost;
         **d)    all of the above.**

18.22    The biological control of the prickly pear in Australia involved the use of:
         a)    a scale insect;
         **b)    a moth;**
         c)    a leaf-eating beetle;
         d)    blowfly larva.

18.23    Screwworms were controlled by raising factory-reared sterile:
        **a)**    **males;**
        b)    females;
        c)    eggs;
        d)    pupae.

18.24    Which of the following is not applicable to population harvesting?
        **a)**    **to obtain a croppable surplus the population must be allowed to reach its maximum density;**
        b)    though there are two levels of density from which a given sustained yield can be obtained, the maximum sustained yield can only be harvested at one density;
        c)    if some of the standing crop is removed each year the population will decline, but will stabilize at a level at equilibrium with the rate of harvesting;
        d)    all of the above are correct answers.

18.25    Wild turkeys were restored by:
        a)    genetically breeding an adaptable turkey;
        b)    eliminating all the turkey predators;
        **c)**    **transporting small groups of turkeys from areas of abundance to ideal habitats elsewhere;**
        d)    declaring the turkey a nationally protected species.

**TRUE/FALSE**

18.26  **T**  F    Most pest and weed populations show a mixture of r and K characteristics.

18.27  T  **F**    Synthetic pesticides are usually species specific.

18.28  T  **F**    Organophosphates are water soluble and persist longer in the environment than chlorinated hydrocarbons.

18.29  **T**  F    Since organophosphates are water soluble they often enter and contaminate water sources.

18.30  T  **F**    Screwworm flies were controlled in Texas by using pheromones to lure males to traps.

18.31  T  **F**    The object of IPM is to wait until the control of a pest is absolutely necessary, for example when the pest population is increasing rapidly.

18.32  **T**  F    C.D.J. Petersen developed a method for tagging fish that has helped researchers to estimate the size of commercial fish populations.

18.33  **T**  F    When humans first harvest a population they tend to take the large and old individuals first.

18.34  **T**  F    One way to increase the carrying capacity of a population is to remove several individuals from the population and then stabilize it.

18.35  T  **F**    The blue whale has been forced into near extinction because for the last 40 years exploiters have concentrated on killing mature males and older reproductively active females.

18.36  **T**  F    The most important component of any population exploitation is economics.

18.37  **T**  F    In forests the goal of a sustained yield program is to reach a balance between net growth and harvest.

18.38  T  **F**    Old-growth forests can be cut and if given time the old-growth stage will return.

18.39  T  **F**    Pesticides have been used by human populations for less than 40 years.

18.40  **T**  F    Malathion is an organophosphate.

**MATCHING**

A.   Kirtland warbler
B.   bison
C.   elephant
D.   Arabian oryx
E.   California condor

18.41  __**E**__    rear and breed individuals in captivity

18.42  __**A**__    controlled burning of jack pine forests

18.43  __**D**__    reintroduced animals into suitable habitat

18.44  __**B**__    restrict domestic livestock on their range

18.45  __**C**__    crop surplus animals

## DISCUSSION

18.46    "Economic maturity is different than ecological
         maturity."  Would you agree or disagree with that
         statement?  Give evidence to support your answer.

18.47    What are some environmental dangers that we can expect
         from the use of long-lasting, non-specific
         insecticides?

18.48    What is meant by sustained yield?  What is the
         difference between maximum sustained yield and optimal
         sustained yield?  What are some problems that surround
         the idea of sustained yield?

18.49    What characteristics would you want to incorporate into
         an environmental "safe" insecticide?

18.50    What types of conservation practices could be
         instituted to protect wildlife?  Would you agree or
         disagree with the statement, "Every organism has a
         right to life?"  Explain your answer.

# Chapter 19

## Community Organization and Structure: Spatial Patterns

<u>SENTENCE COMPLETION</u>

19.1    A **community** is a collection of plants and animals interacting in the same environment.

19.2    The form and structure of a community is determined by the nature of the **vegetation.**

19.3    The species that has the greatest influence on regulating or modifying an environment is called the **dominant species.**

19.4    The number of species in a community is a measurement of **species diversity.**

19.5    **Alpha** diversity is a measurement of diversity within a community.

19.6    As the number of microhabitats in an environment **increases** the species diversity increases.

19.7    In a **heterogeneous** environment the structure of the community is complex and the number of niches is many.

19.8    The **predation** theory suggests that as a predator selects its prey it reduces the level of competition among prey species.

19.9    The topmost layer in the vertical structure of a forest is called the **canopy.**

19.10   In a lake the deep, cold layer of dense water near the bottom is called the **hypolimnion.**

19.11   In both land and water environments photosynthesis takes place in the **autotrophic** layer.

19.12   An area where two or more distinct communities merge and intergrade is known as an **ecotone.**

19.13   The **edge effect** is based on the observation that species variety and density is greatest where two communities merge or touch.

19.14   According to zoographer P. Darlington, as the size of an area **increases** the number of species **increases.**

19.15    The rate of extinction of a species would be **greater** on **small** islands than on large ones.

## MULTIPLE CHOICE

19.16    Christen Raunkiaer would classify desert plants as:
   **a)   therophytes;**
   b)   hemicryptophytes;
   c)   phanerophytes;
   d)   chamaephytes.

19.17    The _____ diversity is the diversity between biotic communities.
   a)   gamma;
   **b)   beta;**
   c)   alpha.

19.18    Plant species diversity is influenced by:
   a)   type of soil;
   b)   soil nutrients;
   c)   elevation;
   **d)   all of the above.**

19.19    "The more nutrients there are in a specific ecosystem, the greater the diversity." This statement could be offered to support the _____ hypothesis.
   a)   predation;
   **b)   productivity;**
   c)   climatic;
   d)   heterogeneous.

19.20    Species diversity increases from:
   a)   the equator to the arctic;
   b)   the open ocean to continental shelves;
   **c)   from east to west in North America;**
   d)   from continental shelves to peninsulas.

19.21    The _____ is the primary site of energy production in a forest.
   a)   herb layer;
   b)   understory;
   **c)   canopy;**
   d)   shrub layer.

19.22    Reptiles have their greatest density in _____ of North America.
   **a)   deserts;**
   b)   grasslands;
   c)   chaparral;
   d)   boreal forests.

19.23    From top to bottom, what is the correct sequence of
         layers in a lake?
         a)    hypolimnion, metalimnion, epilimnion;
         **b)    epilimnion, metalimnion, hypolimnion;**
         c)    metalimnion, hypolimnion, epilimnion;
         d)    epilimnion, hypolimnion, metalimnion.

19.24    An inherent edge:
         a)    is maintained by periodic disturbances;
         **b)    is stable and permanent;**
         c)    is successional or developmental;
         d)    is human-induced.

19.25    The autotrophic layer of a lake is the:
         a)    herbaceous layer;
         **b)    epilimnion;**
         c)    hypolimnion;
         d)    thermocline.

**TRUE/FALSE**

19.26   **T**   F    Epiphytes are plants that grow on other plants and
                     have air roots.

19.27   T   **F**    Deserts would be made up of chamaephytes.

19.28   T   **F**    The dominant species of an environment is always
                     one that is most abundant.

19.29   **T**   F    Subdominants are more specialized than dominants
                     and are more specific in physiological tolerances.

19.30   **T**   F    When you examine a community of plants and animals
                     you discover that few species are abundant, in
                     fact most are rare.

19.31   **T**   F    The Shannon index considers the number of species,
                     also the relative abundance of species when
                     finding diversity.

19.32   **T**   F    As you approach the Equator the number of species
                     of plants and animals increases.

19.33   T   **F**    If a forest has a closed canopy then the
                     understory will be well developed.

19.34   T   **F**    The epilimnion is characterized by a thermocline.

19.35   **T**   F    Herbivory, predation, and decomposition occur in
                     the heterotrophic layer of an ecosystem.

19.36  T  **F**  An edge that has been induced by natural disturbances is termed inherent.

19.37  **T**  F  As the number of colonizing species on an island increases the number of new colonizers arriving decreases.

19.38  **T**  F  The minimum size for a forest habitat will depend on the minimum area at which moisture and light conditions are suitable for xeric species to grow.

19.39  T  **F**  The dominant species is the one in the community that either produces the most energy or cycles the most nutrients.

19.40  **T**  F  Mountain area support more species than flatlands.

**MATCHING**

A.   phanerophytes
B.   hemicryptophytes
C.   therophytes
D.   chamaephytes
E.   cryptophytes

19.41  __E__  perennial buds buried in the ground in cold, moist climates.

19.42  __A__  typical of moist, warm environments; trees over 25 cm. tall.

19.43  __B__  typical of cool, moist climates; buds at ground level.

19.44  __D__  perennials growing to 25 cm in cool, dry climates.

19.45  __C__  annuals in deserts and grasslands.

**DISCUSSION**

19.46  What are the major points covered by the island equilibrium model?  Is there evidence to support this model?  What are its weaknesses?

19.47  What is a community?  What are some main characteristics of communities?

19.48  What is species dominance?  How is it decided?

19.49  Why is it difficult for ecologists to distinguish between population ecology and community ecology?

19.50    What are the major components acting on species
         diversity in a community?  What explanations can you
         give to account for the fact that species diversity is
         greater in the tropics than in the arctic regions of
         Earth?

# Chapter 20

## Community Changes:  Temporary Patterns

<u>SENTENCE COMPLETION</u>

20.1    The replacement of one community by another over time
        is **succession.**

20.2    **Primary succession** begins on land that has never been
        colonized before.

20.3    The **facilitation model** for succession suggests that
        organisms in each stage alter the environment and
        prepare the way for organisms of later stages.

20.4    In early stages of succession soils are deficient of
        nutrients, especially **nitrogen.**

20.5    **Plankton** are the first forms of life that colonize a
        pond.

20.6    Succession that takes place after human disturbances is
        **secondary succession.**

20.7    During succession, the number of species increases and
        the community **biomass** increases.

20.8    The greatest diversity of animal species is supported
        by the herb and **shrubland** stages of succession.

20.9    **Cyclic succession** begins when some periodic
        disturbance, such as frequent ground fires, restarts
        succession.

20.10   **Fluctuations** are nonsuccessional changes that occur in
        plant communities.

20.11   The stable, self-perpetuating community at the end of a
        succession is the **climax.**

20.12   The climax species of vegetation are **more** shade-
        tolerant than the early plant colonists.

20.13   Much of the present pattern of Earth's vegetation was
        influenced by climatic events of the **Pleistocene.**

20.14   The food chains in a climax community are **more** complex
        than in the early successional stages.

20.15   The early plant colonists of succession are **r-selected**
        organisms.

20.16    The type of climax community that develops in an area
         depends on:
         a)    the bedrock present;
         b)    the type of early successional phases the
               area experienced;
         **c)    the climate;**
         d)    the nutrients in the soil.

20.17    Which of the following is unlikely to be the site of
         secondary succession?
         a)    an acre of forest cleared to plant a crop;
         **b)    lava crust surrounding a volcano;**
         c)    the road cut along the side of a freeway;
         d)    a pastureland grazed by cows.

20.18    In the early stages of succession:
         a)    soils are ow in nutrients;
         b)    organic matter in the soils is abundant;
         c)    light is abundant;
         **d)    a and c, but not b;**
         e)    a, b, and c.

20.19    An early successional stage would contain which of the
         following?
         **a)    lichens;**
         b)    shrubs;
         c)    trees;
         d)    grasses.

20.20    Compared with an early stage of succession a climax
         community would have _____ ecological niches.
         a)    few;
         **b)    many;**
         c)    simple;
         d)    small.

20.21    A pond in a forest will pass through new stages of
         ecological succession because of:
         a)    algae blooms;
         b)    oxygen depletion;
         **c)    sedimentation;**
         d)    leaching.

20.22    Which statement is correct?
         a)    each successional community is a one-time
               product of a special set of abiotic and
               biotic factors;
         b)    the exact original composition of a community
               can never be duplicated after a disturbance
               has been made;
         c)    you can predict the type of vegetation that
               will appear in a region, but you cannot
               predict the specific type of local
               vegetation;
         **d)    all the statements are correct.**

20.23    Which of the following human activities may disrupt a
         climax community and start up succession?
         a)    grazing;
         b)    lumbering;
         c)    highway construction;
         **d)    all of the above.**

20.24    A successional stage will last:
         a)    one year;
         b)    five years;
         c)    250 years;
         **d)    differs from one ecosystem to another.**

20.25    If global temperatures were to rise two degrees
         Celsius:
         a)    tundra would replace conifers;
         b)    conifers would replace deciduous trees;
         **c)    deciduous trees would replace conifers;**
         d)    boreal plants would replace temperate plants.

**TRUE/FALSE**

20.26    T  **F**    Succession involves the change of species during
                     the seasons of the year.

20.27    **T**  F    Bare rock will first be colonized by lichens.

20.28    T  **F**    The development of new biotic communities on sand
                     dunes is an example of heterotrophic succession.

20.29    T  **F**    The colonizing organisms in succession are K-
                     selected organisms.

20.30    **T**  F    Succession takes place as availability of
                     resources changes with time.

20.31    T  **F**    The plants in the early successional stages are
                     large and slow-growing.

217

20.32  T  F  A marsh is a successional stage of a pond characterized by emergent vegetation.

20.33  T  F  Succession in microcommunities is characterized by maximum availability of energy and nutrients at the start of succession and a steady decline in these resources as succession proceeds.

20.34  T  F  Disturbed ecosystems are most often colonized by weeds.

20.35  T  **F**  Animal ecologists have found characteristic animals life that corresponds to each of the stages in plant succession.

20.36  T  F  Bobwhite quail, cottontail rabbit, and woodcock are all animals that depend on the early stages of succession for survival.

20.37  T  F  Each successional community is a one-time event not to be repeated.

20.38  T  F  Fluctuations are different from successions because during fluctuations no new species invade the area and the dominant species will return in time.

20.39  T  F  Even in a climax community succession never ceases because changes are constantly occurring in some part of the area.

20.40  T  F  Secondary succession is likely to occur on artificially disturbed land.

MATCHING

There are three models used to explain succession:

        A.  Facilitation model
        B.  Tolerance model
        C.  Inhibition model

20.41  __C__  A location is colonized by a species of plants that maintains its position against all invaders.

20.42  __A__  Each successional stage alters the environment allowing later stages to become established.

218

20.43  __B__  Later successional stages become established because they are efficient at exploiting resources.

20.44  __A__  Shade tolerant trees do well under the shrubs of an early stage of succession.

20.45  __C__  The ultimate winners in succession are the long-lived plants.

**DISCUSSION**

20.46  What is the difference between primary and secondary succession?  What is heterotrophic succession?

20.47  What is a climax community?  What factors are important in establishing a climax community?  How long will it take for successional stages to reach a climax stage?

20.48  Succession at any location involves facilitation, tolerance, and inhibitation.  Describe the characteristics of each of these models and describe how they influence the successional outcome.

20.49  What would be the stages of transition that would occur from a pond to a terrestrial community?

20.50  Fires are important to several communities of plants. What role does fires plant during succession?

# Chapter 21

## Natural Disturbance and Human Impact

### SENTENCE COMPLETION

21.1    **Gaps** result from small-scale disturbances that create a site for regeneration and growth

21.2    Landscape diversity could not be maintained without **disturbances.**

21.3    In the western United States **70%** of the forest fires are caused by lightning.

21.4    Fires in **grasslands**, set by humans, have increased over the years.

21.5    **Surface** fires kill herbaceous plants and woody seedlings.

21.6    **Crown** fires are common in coniferous forests.

21.7    The soil would have to warm up to above **100** degrees Celsius before moisture in it would evaporate.

21.8    The **Kirtland warbler** is dependent on regular burning of jack pine to survive.

21.9    **Even-aged** management will reverse the process of succession and will lead to the development of an early stage of succession.

21.10   A type of **clear-cutting** that leaves 10 to 70 percent of the old growth stand for later harvest is shelterwood cutting.

21.11   The development of **agriculture** has radically altered the pattern of vegetation worldwide.

21.12   **Surface** mining accounts for most of the production of coal and iron.

21.13   The tendency of ecosystems to reach a steady state is **stability.**

21.14   **Resilience** is the ability of a system to maintain equilibrium although the environment is changing.

21.15   **Aquatic** ecosystems are less stable than terrestrial ones.

# MULTIPLE CHOICE

21.16    Gaps in a forest:
  a)    decrease light availability;
  b)    increase soil moisture;
  c)    decrease soil nutrients;
  **d)    increase nutrients.**

21.17    What conditions are necessary for fire to be important?
  a)    you need some natural way to start a fire:
  b)    dry-weather;
  c)    sufficient organic matter to burn;
  **d)    all of the above are needed.**

21.18    Which of the following communities is fire dependent?
  **a)    savannas;**
  b)    deserts;
  c)    tundra;
  d)    hardwood forests.

21.19    The most damaging type of fire is a _____fire.
  a)    surface;
  **b)    ground;**
  c)    crown.

21.20    Frequent low-intensity surface fires:
  a)    prevent a buildup of ground litter;
  b)    eliminate shade-tolerant competitive trees;
  c)    destroy soil microorganisms;
  **d)    a, b, but not c;**
  e)    a, b, and c.

21.21    Trees that are dependent on fire:
  a)    grow rapidly;
  b)    have thick bark;
  c)    have cones that open when exposed to flames;
  **d)    all of the above.**

21.22    Which of the following animals is dependent on fire?
  a)    bald eagle;
  **b)    Sylvia warbler;**
  c)    snowy egret;
  d)    wild turkey.

21.23    _____ have defoliated many acres of
forest in the United States.
  a)    screwworms;
  b)    swallowtail butterflies;
  **c)    gypsy moths;**
  d)    grasshoppers.

21.24    Selection cutting:
      a)    is widely practiced in the Pacific Northwest;
      b)    leads to the rapid replacement of herbs, shrubs, and sprout growth.
      **c)    has little effect on forest composition;**
      d)    favors shade-intolerant trees because large stands of forest are removed.

21.25    Most of the timber harvested in the United States involves:
      a)    selection cutting;
      **b)    clear-cutting;**
      c)    shelterwood cutting;
      d)    strip cutting.

**TRUE/FALSE**

21.26    **T**    F    Gap formation is essential in maintaining the diversity of tropical forests.

21.27    **T**    F    After a large-scale disturbance such as a forest fire, the area will be colonized by opportunistic species.

21.28    T    **F**    In a coppice forest, cutting out of the canopy will cause the characteristic woodland species to be permanently replaced by opportunistic annual plants.

21.29    T    **F**    In the western United States most forest fires are started by careless campers.

21.30    **T**    F    Crown fires are common in coniferous forests.

21.31    **T**    F    Jack pine and lodgepole are fire-dependent species.

21.32    T    **F**    Native grasslands in North America experienced ground fires every 100 years.

21.33    **T**    F    Fires could increase the surface runoff of water and lead to soil erosion.

21.34    **T**    F    When elephants exceed their carrying capacity they damage their habitat and convert forests into grasslands.

21.35    **T**    F    Selection cutting maintains the general composition of the forest.

21.36   T   **F**     In the Pacific Northwest loggers selectively cut the forests and reseed with several species of native trees.

21.37   **T**   F     Anthropologists believe that the collapse of many of the world's great civilizations was the result of poor agricultural practices and excessive soil erosion.

21.38   **T**   F     Urban plant communities are composed of small pockets of native vegetation that were saved when green belt areas were established.

21.39   **T**   F     Surface mining is ecologically unsound because it destroys mountains, increases siltation in streams, and releases toxic elements into the environment.

21.40   **T**   F     Plant communities that are the most resistant to change have a large biotic structure and large standing biomass.

## MATCHING

    A.   surface fires
    B.   crown fires
    C.   ground fires

21.41   __B__     essential for the maintenance of jack pines.

21.42   __A__     destroys thin-barked trees, but not thick-barked trees.

21.43   __A__     the most common type of fires that kill herbaceous plants and seedlings.

21.44   __C__     destroys the humus down to the mineral substrate.

21.45   __B__     burns the canopy.

## DISCUSSION

21.46     Several forces act as agents of disturbance. Describe ways in which wind, moving water, and drought may alter the environment.

21.47     What role does fire play in natural ecosystems? How may the prevention of fire lead to problems for a forest?

21.48   What impact did fire have on Yellowstone Park?  Are
        there special adaptations of plants and animals that
        will help them cope with fire?

21.49   Describer the difference between selection cutting and
        clear-cutting.  Which practice is the most ecologically
        sound?  Why?

21.50   What impact has urbanization had on natural biotic
        communities?  Are there urban plans that could be
        designed and incorporated to minimize the effects of
        urbanization on surrounding ecosystems?

# Chapter 22

## Production in Ecosystems

### SENTENCE COMPLETION

22.1    **Potential energy** is energy at rest that is available to do work.

22.2    "Energy cannot be created or destroyed." This is a statement of the **first law of thermodynamics**.

22.3    Photosynthesis is an **endothermic** reaction.

22.4    When energy is passed from one organism to another, part of the energy is lost as **heat**.

22.5    **Gross primary production** is the total amount of energy from the sun trapped by a plant during photosynthesis.

22.6    The amount of accumulated organic matter found in a given area at a specific time is the **standing crop biomass**.

22.7    The water depth at which light intensity is just sufficient to meet the biological needs of the plankton and plants is known as the **compensation level**.

22.8    Most of the light absorbed by plants is not used in photosynthesis but is converted to **heat**.

22.9    In water light intensity **decreases** as depth **increases**.

22.10   The productivity of ecosystems is influenced by changes in **temperature** and **precipitation**.

22.11   Once energy is consumed it is used for **maintenance, growth,** and **reproduction** or it is released as feces or urine.

22.12   The energy available for heterotrophs is the **net** production.

22.13   **Tropical rain forests** and **coral reefs** are the most productive ecosystems on Earth.

22.14   As a forest gets older most its energy production is used for **maintenance** and very little is used for **growth**.

22.15   The energy used for body **maintenance** is lost as **heat**.

## MULTIPLE CHOICE

22.16    Which of the following is abiotic?
        a)   producers;
        b)   algae;
        **c)   sunlight;**
        d)   top carnivores.

22.17    All the following are autotrophs except:
        **a)   a mushroom;**
        b)   a pine tree;
        c)   grasses;
        d)   a shrub.

22.18    Photosynthetic organisms are;
        **a)   autotrophic;**
        b)   heterotrophic;
        c)   consumers;
        d)   decomposers.

22.19    The second law of thermodynamics would explain why:
        a)   there is an increase in energy as you
            progress through a food chain;
        b)   there is more energy in consumers than
            producers;
        c)   there is an increase of energy from
            autotrophs to heterotrophs;
        **d)   there is a decrease in energy as you progress**
            **through a food chain.**

22.20    Most of the energy in an ecosystem is found in the:
        a)   decomposers;
        b)   herbivores;
        **c)   autotrophs;**
        d)   omnivores.

22.21    Plants with a high shoot-root ratio:
        **a)   are well adapted to harsh environments;**
        b)   have most of their biomass above ground;
        c)   would include the plants in a forest;
        d)   have most of their biomass below ground.

22.22    Maximum photosynthesis in a lake takes place:
        a)   at the surface of the water;
        b)   in the middle zone of the lake;
        c)   near the bottom;
        **d)   below the surface where the light is at**
            **saturation.**

22.23    Most of the light absorbed by plants is converted to
_____:
     a)    protein;
     b)    glucose;
     **c)    heat;**
     d)    lipids.

22.24    One of the least productive ecosystems is a:
     a)    boreal forest;
     b)    tropical rain forest;
     c)    grassland;
     **d)    tundra.**

22.25    Poikiotherms:
     a)    have a high assimilation efficiency;
     b)    have low production energy;
     **c)    use most of their energy for maintenance;**
     d)    use most of their energy for growth.

## TRUE/FALSE

22.26    **T**    F    The heterotrophs in an ecosystem would include
insects, birds, and mammals.

22.27    T    **F**    The word ecotone was created to describe an
energy-processing system whose various components
have evolved together over a long period of time.

22.28    **T**    F    Energy exists in two forms, either potential or
kinetic energy.

22.29    **T**    F    According to the first law of thermodynamics
energy cannot be created or destroyed, but it can
be changed from one form to another.

22.30    **T**    F    A reaction that releases energy and produces heat
is exothermic.

22.31    T    **F**    Heat is an important form of energy because it can
be used by cells to do work.

22.32    **T**    F    Life is an open energy system that is maintained
in a steady state.

22.33    T    **F**    Gross primary production is the energy available
to a plant after some is used for respiration and
some is stored as organic matter.

22.34    T    **F**    Biomass is a measurement of the rate at which
organic matter is created by photosynthesis.

22.35　T　F　Annuals first allocate their photosynthate to leaves and then later to flowers.

22.36　T　**F**　If the photosynthate is limited, it is supplied for reproduction first.

22.37　T　**F**　The region of maximum energy productivity in a lake would be in the upper sunlit surface.

22.38　**T**　F　Production efficiency is the ratio of net primary production to gross primary production.

22.39　T　**F**　Woody plants have a higher production efficiency than grasses.

22.40　**T**　F　Homeotherms have a higher assimilation efficiency but a low production efficiency.

## MATCHING

There are three major components of an ecosystem. In a forest these components would be represented by:

A.　producers
B.　consumers
C.　abiotic

22.41　__A__　the understory.

22.42　__C__　the litter on the forest floor.

22.43　__B__　scale insects on the leaves of a shrub.

22.44　__C__　the humidity levels in the canopy at night.

22.45　__B__　the earthworms in the leaf litter.

## DISCUSSION

22.46　The first and second laws of thermodynamics describe many activities if energy in ecosystems. What are these two laws? Give examples of how these laws decide many activities of energy in ecosystems.

22.47　What is production? How does gross primary production differ from net primary production? What is assimilation efficiency and how is it determined? What information would you need to find growth efficiency?

22.48    The pattern of photosynthate allocation varies from one
         type of plant to another.  Describe the patterns of
         allocation for each of the following:

                    a.   planktonic algae;
                    b.   annuals;
                    c.   perennials;
                    d.   woody shrubs;
                    e.   deciduous trees;
                    f.   evergreen trees.

22.49    Productivity differs from one ecosystem to another.
         What factors influence the productivity of ecosystems?
         How does productivity change with time?

22.50    What is secondary production?  What are some of the
         variables that may affect secondary production?  How
         does the ability of consumers to convert the energy it
         ingests differ from one species to another?

# Chapter 23

## Trophic Structure

**SENTENCE COMPLETION**

23.1    The energy in the ecosystem passes through a series of transfers, one population eating another. These transfers comprise a **food chain**.

23.2    Plant eaters are called **herbivores**. Animal eaters are known as **carnivores**.

23.3    The major pathway of energy flow in a terrestrial ecosystem is a **detrital food chain**. The **grazing food chain** is the most common energy flow in an ecosystem with a high rate of harvest and a rapid turnover of organisms.

23.4    The energy available to an organism is either used to **do work** or to build **organic molecules**.

23.5    A carnivore that feeds on herbivores would be called a **second level consumer**.

23.6    The dead organic matter that collects in an ecosystem that is attacked by saprophages is called **detritus**.

23.7    Bacteria that require oxygen are **aerobic**; organisms that can carry on metabolism without oxygen are **anaerobic**.

23.8    About **10%** of the energy captured by organisms at one trophic level is converted to organic molecules at the next highest trophic level.

23.9    About **1%** of the solar energy that falls on a leaf is converted into chemical energy by photosynthesis.

23.10   The rate at which solar energy is converted into chemical energy of organic molecules is the ecosystem's **gross primary productivity**.

23.11   In lakes, the primary production of energy is concentrated in **microscopical algae**, whereas in terrestrial ecosystems the primary production is concentrated in **large** plants with long life-cycles.

23.12   According to the **second law of thermodynamics** no transfer of energy is 100% efficient.

23.13    **Food webs** develop because several food chains interlink
in a complex series of interactions.

23.14    Food chains in **fluctuating environments** will be short
and will have few trophic levels.

23.15    The **net primary productivity** of an ecosystem is the
total amount of producer biomass per area per amount of
time, or the **gross productivity** minus the energy used
in **cellular metabolism**.

## MULTIPLE CHOICE

23.16    The ultimate source of energy for all life is:
      a)    a glucose molecule;
      b)    an amino acid;
      **c)    sunlight;**
      d)    decomposition.

23.17    About _____ of the solar energy that falls on a leaf
is converted into chemical energy.
      a)    10%;
      b)    50%;
      c)    80%;
      **d)    1%.**

23.18    Which of the following is a decomposer?
      a)    a vulture;
      b)    bot flies;
      **c)    soil bacteria;**
      d)    Indian pipe.

23.19    The grazing food chain is the dominant system in:
      a)    grasslands;
      b)    forests;
      c)    wetlands;
      **d)    the open ocean;**
      e)    there is more than one correct answer.

23.20    In a forest detrital food chain up to _____ of the
gross energy production goes into body maintenance and
respiration.
      **a)    50%;**
      b)    20%;
      c)    35%;
      d)    13%.

23.21    In a food web which organism is the least numerous?
         a)    producers;
         **b)    carnivores;**
         c)    herbivores;
         d)    decomposers.

23.22    The first trophic level in a biomass pyramid is made up
         of:
         a)    carnivores;
         b)    omnivores;
         c)    decomposers;
         **d)    producers.**

23.23    The _____ are part of the last trophic level in
         a food pyramid.
         **a)    top carnivores;**
         b)    producers;
         c)    herbivores;
         d)    decomposers.

23.24    _____ productivity is the rate at which
         new tissue is formed in the producers of a food chain.
         **a)    net primary;**
         b)    net secondary;
         c)    gross primary;
         d)    gross secondary.

23.25    If you began with 1000 calories stored in a few blades
         of grass, how many calories would you expect to be
         passed on to the primary consumer?
         a)    10;
         **b)    100;**
         c)    1000;
         d)    1;

**TRUE/FALSE**

23.26  **T**  F    Herbivores are adapted to eat foods high in
                   cellulose.

23.27  T  **F**    Energy and nutrients enter a food chain by way of
                   the scavengers.

23.28  **T**  F    Since predators are organisms that feed on the
                   tissue of another organism, parasites could be
                   considered predators.

23.29  T  **F**    Most shallow-water ecosystems would be
                   characterized by grazing food chains.

23.30  **T**  F   In terrestrial ecosystems little of the primary production is passed onto herbivores, most enters a detrital food chain.

23.31  **T**  F   Millipedes, earthworms, and nematodes are detritus feeders.

23.32  T  **F**   Saprophages are macrofauna, that is they are big enough to see with the naked eye.

23.33  **T**  F   Detritus is attacked in an orderly fashion by saprophages.  First the sugar-consuming bacteria attack the glucose molecule, then other bacteria and fungi digest cellulose.

23.34  **T**  F   Mineralization is a process that makes soluble nutrients extracted from decomposing matter available to plants.

23.35  T  **F**   Bacterial action tends to release nutrients to the environment during decomposition.  Algae and zooplankton concentrate nutrients.

23.36  **T**  F   Decomposition in tropical rain forests is so rapid that organic material may broken down in one year.

23.37  T  **F**   All organisms, including decomposers belong to specific trophic levels in a food chain.

23.38  T  **F**   If a cow consumed 1000 kilocalories of plant energy about 10 kilocalories would be converted into herbivore tissue.

23.39  **T**  F   Because the biomass of producers must be greater than the biomass of the herbivores they support, it is possible to construct a pyramid of biomass.

23.40  **T**  F   In a pyramid of numbers the trophic level at the base of the pyramid represents the abundance of producers.

**MATCHING**

        A.   omnivores
        B.   scavengers
        C.   saprophytes
        D.   decomposers
        E.   biophages

23.41  __E__   Any organism that uses living material as its source of energy and nutrients.

23.42   __A__   Animals that eat both plants and animals.

23.43   __D__   Bacteria and fungi that convert organic molecules
into inorganic nutrients.

23.44   __B__   Animals that eat dead plants and animals.

23.45   __C__   Plants that are scavenger-like.

## DISCUSSION

23.46   Why is there always a loss of energy with each
progressive step in a food chain?

23.47   What is an ecological pyramid?  What is the difference
between a biomass pyramid and an energy pyramid?

23.48   Of the three types of pyramids--biomass, energy flow
and numbers--only the energy flow pyramid takes the
shape of a real pyramid.  Why is this so?

23.49   What factors decide the size and complexity of food
webs in different ecosystems?

23.50   What are the major differences between grazing food
chains and detrital food chains?  What role do
decomposers play in food chains and food webs?

# Chapter 24

## Cycles in Ecosystems

<u>SENTENCE COMPLETION</u>

24.1   Iron, copper and zinc are examples of **micronutrients** because they are needed by organisms in minute amounts.

24.2   The flow of materials between the abiotic and the biotic parts of an ecosystem is known as a **biogeochemical cycle.**

24.3   Gaseous cycles have the main reservoir of nutrients in the **atmosphere** or **ocean.**

24.4   The force that drives the water cycle is the **sun.**

24.5   **Evaporation** is taking place when more water molecules leave a surface than enter it.

24.6   The three most important gases in the atmosphere are **oxygen, nitrogen and carbon dioxide.**

24.7   Only **3%** of Earth's water supply is available for human use.

24.8   **Oxygen** is a byproduct of photosynthesis.  Its primary role in cells is to act as a **hydrogen** acceptor.

24.9   The biological fixation of nitrogen is accomplished by symbiotic bacteria, by **free-living aerobic bacteria** and by **blue-green algae.**

24.10  The source of all the carbon in living organisms and fossil deposits is **carbon dioxide.**

24.11  During **ammonification** amino acids are broken down by decomposers.

24.12  **Nitrification** is a biological process in which ammonia is oxidized to nitrate and nitrite.

24.13  The **phosphorus** cycle is an example of a sedimentary cycle.  The main reservoirs for phosphorus are **rock** and **phosphate** deposits.

24.14  The **sulfur** cycle is both sedimentary and gaseous.

24.15  Sedimentary cycles involve two phases, **salt solution** and **rock.**

24.16    That portion of the earth that acts as a storehouse for
         an essential element is the:
         a)    exchange pool;
         b)    biotic community;
         **c)    reservoir;**
         d)    all of the above.

24.17    Nitrogen gas makes up approximately _____ of the
         atmosphere.
         a)    50%;
         **b)    78%;**
         c)    2%;
         d)    .03%.

24.18    _____ can incorporate nitrogen gas directly from
         the atmosphere.
         **a)    Bacteria;**
         b)    Fungi;
         c)    Plants;
         d)    Decomposers.

24.19    Which of the following is a micronutrient?
         a)    calcium;
         b)    magnesium;
         c)    sulfur;
         **d)    silicon.**

24.20    Which of the following is a sedimentary biogeochemical
         cycle?
         **a)    phosphorus cycle;**
         b)    carbon cycle;
         c)    nitrogen cycle;
         d)    water cycle.

24.21    The atmosphere holds about a(n) _____ day supply of
         rainfall.
         a)    3;
         b)    50;
         **c)    11;**
         d)    100.

24.22    The major nonliving pool for oxygen include all the
         following except:
         **a)    ozone;**
         b)    molecular oxygen;
         c)    water;
         d)    carbon dioxide.

24.23    All of the following can fix nitrogen from the
         atmosphere except:
         a)    Azotobacter;
         b)    Clostridium;
         c)    Nostoc;
         d)    **Pseudomonas.**

24.24    The least active reservoir for carbon is:
         a)    oceanic;
         b)    terrestrial;
         **c)    atmospheric;**
         d)    geological.

24.25    The process by which nitrogen in nitrates is reduced to
         the gaseous form is called:
         a)    ammonification;
         b)    nitrification;
         **c)    denitrification;**
         d)    fixation.

## TRUE/FALSE

**For the following questions imagine that you are conducting a
study of a lake ecosystem.**

24.26  T  **F**    Phytoplankton in the lake can incorporate nitrogen
                   gas into organic compounds.

24.27  **T**  F    Cyanobacteria in the water reduce nitrogen gas to
                   ammonium.

24.28  T  **F**    Denitrification, the conversion of nitrate to
                   nitrogen gas, will not occur in the lake because
                   this process is limited to terrestrial ecosystems.

24.29  T  **F**    Fish in the lake make use of carbon dioxide to
                   produce complex organic molecules.

24.30  **T**  F    Decomposers on the lake bottom will help release
                   the carbon contained in animals wastes and
                   detritus.

24.31  T  **F**    The concentration of carbon dioxide in the lake
                   will reach its lowest levels at daylight because
                   this is the time the plants respire and
                   photosynthesis ceases.

24.32  T  **F**    Plants produce oxygen during photosynthesis and
                   consume carbon dioxide during respiration.

24.33  T  **F**    In the mud of a lake there is a large reservoir of
                   nitrogen trapped in soil humus.

24.34  T  F    If the levels of phosphorus were suddenly
               increased, the lake would experience an algae
               bloom.

24.35  T  F    It is possible that if the lake is very deep that
               the phosphorus in the surface waters may become
               depleted while the deep water becomes saturated.

24.36  T  **F**    Phosphorus is precipitated in water as calcium
               phosphate that accumulates in the photosynthetic
               zone of the lake.

24.37  **T**  F    Bacteria in the lake will release sulfur, hydrogen
               sulfide and sulfate from organic matter.

24.38  **T**  F    Green and purple bacteria may use hydrogen sulfide
               as an oxygen acceptor in the photosynthetic
               reduction of carbon dioxide.

24.39  T  **F**    If sulfuric acid is added to the lake it will
               quickly be degraded by anaerobic bacteria.

24.40  **T**  F    The movement of nutrients out of the lake
               ecosystem is solved by the fact that plant
               populations will reproduce rapidly when conditions
               are optimal and will take up the available
               nutrients quickly.

## MATCHING

A.  carbon
B.  nitrogen
C.  magnesium
D.  oxygen
E.  phosphorus

24.41  __A__   the major element in all organic molecules.

24.42  __E__   plays a major role in energy transfer in cells.

24.43  __D__   its major reservoir is the atmosphere.  It is
               essential for oxidation reactions.

24.44  __B__   major element in all protein molecules.

24.45  __C__   integral element in the chlorophyll molecule.

## DISCUSSION

24.46  What is the Gaia Theory of biogeochemical homeostasis?
       What may be the possible control mechanisms for this
       homeostasis?

24.47     Are there daily and seasonal patterns that influence
          the concentration of carbon dioxide in the atmosphere?

24.48     What types of mechanisms work in ecosystems to conserve
          and recycle nutrients?

24.49     What impact could human activities have on the dynamics
          of biogeochemical cycles?

24.50     What are the differences between macroelements and
          microelements?  Give examples of each and explain how
          each element is important to a biological activity.

# Chapter 25

## Human Intrusions upon Ecological Cycles

### SENTENCE COMPLETION

25.1    **Unconfined** aquifers are located in porous water-bearing layers of underground rocks.

25.2    Artesian wells are examples of **confined** aquifers, water trapped between layers of impenetrable rocks.

25.3    **Wetlands** are important ecosystems because they hold water that could flood and they degrade toxic wastes.

25.4    The human-made sulfur dioxide in the atmosphere comes from the burning of **fossil fuels** and **coal**.

25.5    The major sources of nitrogen pollution are **agriculture**, **industry**, and **automobile exhaust**.

25.6    Increased levels of nitrogen have been linked to damage of **coniferous** forests.

25.7    **Ozone** is an atmospheric gas that shields the earth from the harmful effects of **ultraviolet light**.

25.8    **Chlorofluorocarbons** have been linked to damage to the ozone layer of the atmosphere.

25.9    **Acid rain** involves the combination of sulfur dioxide, nitrogen oxides, and hydrogen chloride with water vapor and oxygen to form nitric acid and sulfuric acid.

25.10   The intake of **lead** may result in mental retardation, paralysis, hearing loss and death.

25.11   **Chlorinated hydrocarbons** are soluble in lipids and therefore they tend to accumulate in plant and animal tissue.

25.12   Gasoline burning has contributed to significant amounts of **lead** in the atmosphere.

25.13   Insecticides applied by aerial spraying are **dispersed** by the atmosphere, but are **concentrated** by biological activities.

25.14   **DDT** can block calcium metabolism in birds.

25.15    Of the 500 million kilograms of pesticides used
         annually in the United States about 50 percent are
         **herbicides.**

## MULTIPLE CHOICE

25.16    Aquifers are:
         **a)    porous water-bearing layers of underground
                 rock;**
         b)    areas of streams that often flood;
         c)    the watershed near a lake;
         d)    water pollutants that have been associated
               with algae blooms.

25.17    Wetlands:
         a)    help reduce flooding;
         b)    dilute toxic wastes;
         c)    hold water and increase infiltration into the
               ground;
         **d)    all of the above.**

25.18    Photochemical smog is associated with:
         a)    industrial effluent;
         **b)    automobile exhaust;**
         c)    power plants;
         d)    the breakdown of the ozone layer.

25.19    Of all the atmospheric gases _____ is most closely
         associated with air pollution.
         a)    nitrous oxide;
         b)    ozone;
         c)    carbon dioxide;
         **d)    sulfur dioxide.**

25.20    Which of the following can be toxic to organisms?
         a)    cadmium;
         b)    selenium;
         c)    chromium;
         **d)    all of the above.**

25.21    What does Donora, Pennsylvania and the Meuse Valley in
         Belgium have in common?
         a)    the two areas have experienced severe water
               shortages as a result of a drop in the water
               table;
         b)    mercury contaminated streams in both areas
               have caused many illnesses;
         **c)    both areas have experienced many deaths
                 associated with air pollution;**
         d)    insecticides have been found in the food
               supplies of both areas.

25.22    DDT is:
          a)    not soluble in lipids;
          **b)    highly soluble in lipids;**
          c)    highly soluble in water;
          d)    highly soluble in alcohol.

25.23    Highly concentrated amounts of pesticides are used:
          a)    on agricultural crops;
          b)    in Mexico and South America;
          **c)    around the American home;**
          d)    in national forests.

25.24    Which of the following statements about pesticides is true?
          a)    pesticides can interfere with nutrient cycling;
          b)    pesticides can change the population numbers of some organisms;
          c)    pesticides can increase the nitrogen content of some plants;
          **d)    all of the above.**

25.25    The name "Rachel Carson" is closely associated with:
          a)    air pollution;
          b)    acid rains;
          **c)    insecticides;**
          d)    radioactive wastes.

**TRUE/FALSE**

25.26  T  **F**    The southwest part of the United States have been seriously damaged by acid rain.

25.27  T  F    Of the two types of pollutants, primary and secondary, secondary is the more damaging.

25.28  T  **F**    Some areas of Earth have severe water shortages because very little of the water used by humans returns to the water cycle.

25.29  T  **F**    A suburban lawn has a water infiltration rate much higher than that of undisturbed soil.

25.30  **T**  F    Once groundwater is removed the surface of the land often subsides.

25.31  **T**  F    Over one-half of the coastal and interior wetlands of the United States have been drained.

25.32  T  **F**    The contamination of the atmosphere with sulfur dioxide has been greatly reduced by the construction of tall smoke stacks at factories.

25.33　**T**　F　Sulfur dioxide is a major atmospheric pollutant as well as the main cause of acid rain.

25.34　T　**F**　The amounts of nitrous oxides in the atmosphere have been steadily decreasing in the last 30 years.

25.35　**T**　F　The interaction of automobile and industrial pollutants with ultraviolet light results in several secondary pollutants.

25.36　**T**　F　Peroxacetyl nitrate and peroxyproprionyl nitrate are created when nitrogen oxides and volatile hydrocarbons react with oxygen in the presence of sunlight.

25.37　**T**　F　Over short periods of time acid rains can have a beneficial fertilizing effect on plants.

25.38　T　**F**　Acid rains burn vegetation and kill plants on contact.

25.39　T　**F**　The disposal of mercurial wastes into the waters of Minamata Bay, Japan poisoned large birds of prey and weakened the eggs of sea birds.

25.40　T　**F**　DDT does not pose an environmental problem today because it has been banned throughout the world.

**MATCHING**

      A.　lead
      B.　ozone
      C.　DDT
      D.　nitrous oxides
      E.　sulfur dioxide

25.41　__C__　blocks ion transport by inhibiting ATPase.

25.42　__A__　its concentrations are very high in the food chains along roadsides.

25.43　__B__　blocks ultraviolet light when it is in the stratosphere, but close to the ground it is classified as an atmospheric pollutant.

25.44　__E__　its major source comes from the combustion of fossil fuels and coal.

25.45　__D__　a major ingredient in photochemical smog.

## DISCUSSION

25.46     What are the differences between primary and secondary
          pollutants?  Which of the two is more damaging to the
          environment?  Why?

25.47     What is photochemical smog?  How is it formed?

25.48     Ozone shields Earth's surface from the harmful effect
          of ultraviolet light, yet ozone also can be regarded as
          an atmospheric pollutant.  Why?

25.49     How is acid rain formed?  What are some effects of acid
          rain on vegetation, public health, and buildings?

25.50     What are chlorinated hydrocarbons?  What happens to
          them in water and in the atmosphere?  What is their
          biological activity?  Can we expect a decline in the
          levels of DDT in the environment now that the United
          States has banned its use.

# Chapter 26

## Grasslands and Savannas

### SENTENCE COMPLETION

26.1    The **Holdridge life zone system** assumes that the type of vegetation growing in an area is determined by the interaction of temperature and rainfall.

26.2    The distinctive habitat of an area and the community of plants and animals that have evolved in the area form a unit called an **association**.

26.3    Grasses are well adapted to handle the pressures of **grazing** and **fire**.

26.4    The North American grasslands consisted of three types: **tall-grass prairies, short-grass plains**, and **desert grasslands**.

26.5    Annual grasslands are common in California where there are two weather seasons: **rainy** winters and **dry** summers.

26.6    The grasslands in North America have an **east-west** zonation, whereas the Eurasian grasslands have a **north-south** zonation.

26.7    The **ground** layer and the **below-ground root** layer are the two major strata in grasslands.

26.8    Grassland vertebrates include grazing **ungulates** and burrowing **mammals**.

26.9    Most of the primary production of grasslands goes to decomposers of which **fungi** are the largest group.

26.10   A community of plants and animals that is composed of grasses and woody vegetation is known as a **savanna**.

### MULTIPLE CHOICE

26.11   It is believed that grasslands once covered ____of the land surface of Earth.
        a)    12%;
        **b)    42%;**
        c)    50%;
        d)    5%;
        e)    75%.

26.12    All grasslands are characterized by:
         a)    rainfall greater than 800 mm;
         b)    low rate of evaporation;
         c)    woody vegetation;
         **d)    periodic ground fires.**

26.13    Tall-grass prairies:
         **a)    formed a narrow belt running north and south**
         **to the edge of deciduous forests;**
         b)    grade into deserts in the southern United
               States;
         c)    are composed of bunch grasses interspersed
               with mesquite;
         d)    are confined to areas with a Mediterranean-
               type climate.

26.14    Which of the following statements about grasslands is
         false?
         a)    the vegetation is transitory; it grows in the
               spring and dies back in the autumn;
         b)    as the plants die a thick layer of mulch
               accumulates;
         c)    the root layer is more highly developed in
               grasslands than in any other plant community;
         **d)    grassland vegetation is poorly adapted to**
         **fires and is easily destroyed when one passes**
         **through it.**

26.15    Which of the following statements about animal life in
         grasslands is true?
         a)    there are few insect populations in
               grasslands;
         b)    small grazing ungulates are the major
               vertebrates in grasslands;
         c)    the grasslands of Australia are inhabited by
               placental mammals;
         **d)    grazing animals recycle nutrients in grasses**
         **when they produce urine and dung.**

26.16    Grasslands have evolved under grazing pressure.  Which
         of the following will occur to grassland plants when
         they are grazed?
         a)    grasses respond by increasing their
               photosynthetic rate in the remaining tissue;
         b)    grasses increase the total net primary
               production below the ground;
         c)    new growth is stimulated by the increased of
               light intensity near the ground;
         d)    the species composition of the grasses will
               change;
         **e)    there is more than one correct answer.**

26.17    Human activities have had which of the following
         effects on grasslands?
              a)    highly productive native species have been
                    replaced by less productive forage species;
              b)    vegetation cover has been increased;
              c)    **new grasslands have been created by clearing
                    forests;**
              d)    in desert grasslands of North America
                    mesquite density has declined.

26.18    Savannas:
              a)    occur on old alluvial plains;
              b)    **have soil rich in nutrients;**
              c)    are associated with a cool continental
                    climate;
              d)    are composed of woody vegetation sensitive to
                    fire.

26.19    The animal life in savannas:
              a)    include many invertebrates that are the
                    dominant herbivores;
              b)    has stimulated the development of structural
                    and chemical defenses in the plants in
                    response to grazing;
              c)    rarely include ungulates;
              d)    **a and b, but not c;**
              e)    a, b, and c.

**TRUE/FALSE**

26.21    **T**    F    Grazing will stimulate the primary production of
                       grasslands.

26.22    T    **F**    Velds are grasslands associated with South
                       America.

26.23    **T**    F    The density of the woody vegetation found in a
                       savanna is regulated by the amount and
                       distribution of rainfall and the characteristics
                       of the soil.

26.24    **T**    F    Since fires are a regular occurrence in savannas
                       the vegetation must be adapted to fire.

26.25  T  F    The dominant herbivores in a savanna are insects.

26.26  T  **F**    You would expect grassland plants that are palatable to have high concentrations of carbohydrates in their roots so that they can quickly recover from defoliation.

26.27  T  F    Much of the savanna in Africa has been replaced by deserts.

26.28  T  **F**    The shortgrass plains developed in areas of North America where there was abundant and frequent rainfall.

26.29  T  F    Natural grasslands are found in areas where the annual rainfall is between 250 mm. and 800 mm.

26.30  T  F    Soil erosion in grasslands is the result of overgrazing that crops the vegetation and exposes the ground to sunlight, water,and wind.

## MATCHING

**Ecologists recognize that the vegetation of an area is determined by an interaction of temperature and rainfall.  Match the abiotic condition with the expected vegetation type.**

        A.   tundra
        B.   tropical rain forest
        C.   savanna
        D.   grasslands
        E.   desert

26.31   __E__    high temperature, low precipitation.

26.32   __B__    high temperature, high precipitation.

26.33   __A__    cold temperature, low precipitation.

26.34   __C__    high temperature, moderate precipitation.

26.35   __D__    moderate temperature, moderate precipitation.

## DISCUSSION

26.36    In abiotic terms describe the conditions that will lead to the development of a grassland.  What specific adaptation would you expect to find in grassland plants that would allow them to survive under each abiotic condition you listed?

26.37    What are some characteristics of grassland animals that allow them to cope with the unique conditions of this ecosystem?

26.38    How are grasslands and savannas similar?  How are they different?

26.39    What human activities do you think took place in North American grasslands that may have contributed to the Dust Bowl?

26.40    What are biogeographical realms?  Life zones?  Biomes? Explain how climate, soil type, and fire can influence the vegetation type that develops in a specific location.

# Chapter 27

## Shrublands and Deserts

### SENTENCE COMPLETION

27.1    **Shrublands** are seral plant communities, a stage in the environment's move back to a **climax** community.

27.2    **Allelopathy** refers to the release of toxins by shrubs that inhibit the development of their seeds.

27.3    Mediterranean climate is characterized by **hot**, dry summers and **cool**, moist winters.

27.4    Garique, chaparral, mallee, and fynbos are plant communities made up of xeric broadleaf evergreen shrubs called **sclerophylls**.

27.5    The **Northern desert scrub community** is made up of halophytes and C4 plants adapted to environments with warm summers and prolonged cold winters.

27.6    **Precipitation, temperature, soil moisture,** and **nutrients** are all major factors that influence the development of mediterranean-type vegetation.

27.7    If evaporation exceeds rainfall a **desert** community develops.

27.8    Most deserts develop between **15 degrees** and **35 degrees** North and South latitude.

27.9    Deserts often develop on the **lee** side of a mountain because of the **rain shadow effect**.

27.10   Deserts experience a wide daily range in temperatures, **hot** by day and **cool** by night.

27.11   **Estivation** is an lifestyle that allows animals to survive through a dry season.

27.12   **Drought evaders** will flower only when the proper amount of moisture is available.

27.13   In desert plants, the **above-ground shoot** biomass is greater than the root biomass.

27.14   The plant nutrients, **phosphorus** and **nitrogen** are scarce for desert plants.

27.15    Most of the desert community animals are **omnivores,**
         than strict herbivores or carnivores.

## MULTIPLE CHOICE

27.16    Temperatures in deserts drop at night because:
         a)    deserts are found at high elevations;
         b)    high pressure air cells are located over
               deserts;
         c)    desert soils absorb much heat during the day;
         **d)    the air is dry, and without cloud cover, heat
               is not radiated back to the earth's surface.**

27.17    A unique type of vegetation found in California and the
         Mediterranean region is called:
         **a)    chaparral;**
         b)    garigue;
         c)    bajadas;
         d)    arroyos.

27.18    Sclerophylls have:
         a)    small leaves;
         b)    thick cuticles;
         c)    glandular hairs;
         d)    sunken stomata;
         **e)    all of the above.**

27.19    Evidence suggests that the vegetation in mediterranean-
         type ecosystems evolved from _____ flora.
         **a)    tropical;**
         b)    tundra;
         c)    grassland;
         d)    taiga.

27.20    Low growing Eucalyptus is the dominant vegetation type
         in:
         a)    chaparral;
         b)    northern desert scrub;
         **c)    mallee;**
         d)    taiga.

27.21    Mediterranean-type shrublands:
         a)    have a thick understory;
         b)    have a thick ground litter;
         **c)    are highly flammable;**
         d)    are composed of halophytes.

27.22    Which of the following events will not lead to the formation of a desert?
        a)    rainshadow effect of a mountain;
        b)    high pressure cells that alter air movement;
        c)    cold air passing over cold water;
        **d)    strong low pressure cells.**

27.23    Which is not an adaptation found in desert plants?
        a)    hard-coated seeds;
        b)    flowering shortly after a rain;
        c)    water-storage tissue;
        d)    dormancy during the dry season;
        **e)    all of the above are adaptations found in desert plants.**

27.24    Nitrogen fixation in desert soil is accomplished by:
        a)    viruses;
        b)    legumes;
        c)    nitrogen-fixing bacteria;
        **d)    blue-green algae.**

27.25    The animals in deserts are:
        a)    usually large carnivores;
        b)    few in numbers;
        **c)    generalists and opportunists;**
        d)    highly specialized in their feeding behavior.

**TRUE/FALSE**

27.26    **T**    F    Mismanagement of the semiarid regions surrounding the natural deserts of the world has created new deserts.

27.27    **T**    F    Desertification refers to the creation of new deserts throughout the world.

27.28    T    **F**    Shrubs are easy for a taxonomist to classify because they represent a unique evolutionary category of plants.

27.29    T    **F**    Sclerophylls evolved from polar floras and developed in areas with wet, cool summers.

27.30    T    **F**    In North America the xeric broadleaf community of plants is called garigue.

27.31    **T**    F    Prior to human settlement the Mediterranean and California plant communities were dominated by oaks.

27.32   T  **F**      Shrublands have a thick understory and a rich ground litter.

27.33   **T**  F      The cool to cold climatic regions of northwestern Europe were once covered by plants dominated by member of Ericaceae.

27.34   **T**  F      Some shrubland plants avoid droughts by shedding their leaves during the summer.

27.35   T  **F**      The soils of mediterranean-type ecosystems are high in nitrogen and phosphorus.

27.36   **T**  F      High-pressure weather cells may create deserts by deflecting storms and blocking rain from falling in a specific location.

27.37   T  **F**      Many desert animals estivate during the winter months and thus avoid the drop in desert temperatures during that time of year.

27.38   T  **F**      The annual primary production of desert vegetation is higher than that of grassland vegetation.

27.39   **T**  F      Desert plants retain nitrogen and phosphorus in their stem and have in this way adapted to the short supply of nutrients in desert soils.

27.40   T  **F**      Small desert herbivores are granivores, since they feed on dead leaf litter and lichens.

## MATCHING

Match the shrubland type with the correct description.

         A.    garigue
         B.    mallee
         C.    heathland
         D.    chaparral
         E.    northern desert scrub

27.41   __**B**__      An Australian shrub community dominated by <u>Eucalyptus</u>.

27.42   __**D**__      A California mediterranean plant community dominated by oak.

27.43   __**A**__      A mediterranean region vegetation type which results from the degradation of pine forests.

27.44 __C__ A community of plants, including evergreen
sclerophylls, hemicryptophytes, and therophytes.

27.45 __E__ A community characterized by warm summers and C4
plant species, and halophytes adapted to saline
soils.

## DISCUSSION

27.46 What is a shrub? Why is it difficult to develop a
general description that will characterize all shrubs?

27.47 What set of climatic conditions will lead to the
development of a shrubland?

27.48 Precipitation, temperature, soil, moisture, and
nutrient availability all influence the function of
shrublands. How are the plants in a shrubland adapted
to each of these environmental influences?

27.49 What impact has human activities had on the shrublands
and deserts of the world?

27.50 What special abiotic conditions lead to the development
of a desert? How do desert animals cope with the arid
conditions of deserts?

# Chapter 28

## Tundra and Taiga

## SENTENCE COMPLETION

28.1    **Tundra** is a biome characterized by low temperatures, short growing season, and low precipitation.

28.2    Tundra has permanently frozen soil called **permafrost**.

28.3    **Cryoplanation**, the molding of the tundra landscape by frost action is more important than erosion in shaping the arctic terrain.

28.4    **Alpine tundra** lacks a permafrost and is limited to the high elevations of mountains.

28.5    **Tropical alpine tundra** is characterized by having a 24-hour freeze-thaw cycle.

28.6    The **lemming** is a small tundra herbivore that exhibits a three-to-four year density cycle.

28.7    The primary production of the tundra is **low**, because temperature is **low**, growing seasons are **short**, and nutrients are **scarce**.

28.8    Alpine tundra is **more** productive than arctic tundra.

28.9    The two major nitrogen sources for tundra plants are **precipitation** and **biological fixation**.

28.10   Forests at high altitudes consist of low growing, wind-shaped trees.  This transition area between forest and tundra is called **Krummholz**.

28.11   **Taiga** is the largest vegetation community covering 11% of Earth's land surface.

28.12   Conifers and broadleaf trees are often subjected to periodic **fires** that provide suitable habitat for seed germination and reduces competition from **hardwood** tree species.

28.13   The **caribou** is the major grazing animal on the tundra.

28.14   In the taiga, mosses growing on the ground **lower** soil temperature and **increase** soil moisture, thereby speeding up/**slowing down** the rate of decomposition.

28.15    The dominant tree species in a boreal forest are
         **spruces** and **pine**, but successional stages may include
         **birch** and **poplar**.

## MULTIPLE CHOICE

28.16    Tundra is characterized by:
         a)    low temperatures;
         b)    short growing season;
         c)    high precipitation;
         **d)    a and b, but not c;**
         e)    a, b, and c.

28.17    Permafrost:
         a)    is a characteristic of alpine tundra soil,
               but not of arctic tundra soil;
         b)    traps water deep within the soil and makes
               the top layers very dry;
         **c)    limits the depth to which roots can grow;**
         d)    encourages microorganism to decompose the
               humus.

28.18    Vegetation in the tundra:
         a)    consists of many different species of plants;
         b)    shows rapid growth during all yearly seasons;
         **c)    propagates itself almost entirely by
               vegetative means;**
         d)    has surface hairs on its leaves that help the
               plant lose metabolic heat.

28.19    Which abiotic factor is more likely to have an
         influence on alpine tundra plants than on arctic tundra
         plants?
         a)    low temperature;
         **b)    high ultraviolet light intensity;**
         c)    high temperatures;
         d)    permafrost.

28.20    The cushion-like growth pattern of alpine tundra plants
         is an adaptation for:
         a)    nutrient conservation;
         **b)    heat conservation;**
         c)    water conservation;
         d)    light conservation.

28.21     Tropical tundra has:
          a)   little seasonal variation in mean daily
               temperature;
          b)   little seasonal variation in rainfall;
          c)   great daily variation in temperature;
          d)   a and b;
          **e)   a and c, but not b.**

28.22     The fact that the growth forms and physiology of many
          tropical alpine tundra plants is similar although they
          belong to different species and genera is evidence of:
          **a)   convergent evolution;**
          b)   coevolution;
          c)   divergent evolution;
          d)   parallel evolution.

28.23     Which of the following is not an adaptation you would
          expect to find in tundra plants?
          a)   well developed water-storing pith in the
               xylem;
          **b)   fire retardant seeds;**
          c)   short-lived adventitious roots;
          d)   high leaf index.

28.24     Which statement is false?
          **a)   the species diversity of tundra animals is
               high;**
          b)   there are abundant populations of insects in
               the tundra;
          c)   the dominant vertebrates are herbivores;
          d)   <u>Canis</u> <u>lupis</u> is the major arctic carnivore.

28.25     Tundra soil does not store nutrients well so the plants
          must depend on_____ to make nutrients available to
          them.
          a)   leaching;
          b)   water percolation;
          c)   legumes;
          **d)   decomposition.**

**TRUE/FALSE**

28.26  **T**  F    Permafrost is impervious to water, and as a
                   result, tundra soils stay soggy.

28.27  T  **F**    Soil microorganisms are very active in tundra soil
                   and decomposition proceeds rapidly.

28.28  **T**  F    Alpine tundra lacks a permafrost.

28.29  T  **F**  Tundra does not possess a vegetation type unique to itself.

28.30  **T**  F  Taiga ecosystems are dominated by softwood forests.

28.31  T  **F**  The tundra temperatures are low therefore very few insects are found in this ecosystem.

28.32  **T**  F  Arctic plants make use of the long summer days by photosynthesizing during the full 24-hr. photoperiod.

28.33  T  **F**  Because tundra is remote and cold it not been disturbed by humans.

28.34  **T**  F  You would expect to find stands of lichens and black spruce in the boreal-mixed forest ecotone.

28.35  **T**  F  Wind and snow may shape the trees in a Krummholz.

28.36  T  **F**  Boreal forests are restricted to a band of vegetation running 30 degrees N and 30 degrees S latitude.

28.37  **T**  F  Plants growing in the North American taiga have low nutrient requirements and must be able to tolerate wet soils.

28.38  **T**  F  The nutrient and energy turnover among the plant communities of a boreal forest is slow.

28.39  **T**  F  Most tundra plants reproduce asexually.

28.40  **T**  F  You would expect to find a Krummholz in environments were temperature is low and winds are strong.

## MATCHING

A.  arctic tundra
B.  alpine tundra
C.  Krummholz
D.  tropical alpine tundra
E.  taiga

28.41  __E__  spruces, firs, pines, grouse, moose.

28.42  __A__  lichens, mosses, caribou, lemmings.

28.43  __C__  flagged trees, dwarf mountain pine, juniper.

28.44   __B__   sedges, heaths, pikas, marmots.

28.45   __D__   tussock grasses, small-leaved shrubs, tree-like
                rosette plants.

## DISCUSSION

28.46   Describe the physical features of the tundra biome?

28.47   Give examples of how plants and animals have adapted to
        the specific conditions of the tundra biome.

28.48   Why is the tundra sometimes called an arctic desert?

28.49   Why is the tundra considered a very fragile
        environment?  What are some examples of human
        activities that have affected tundra?

28.50   What specific abiotic factors are necessary to have a
        taiga?  What are some special adaptations you would
        expect to find among the plants and animals that occupy
        this ecosystem?

# Chapter 29

## Temperate Forests

<u>SENTENCE COMPLETION</u>

29.1    **Deciduous** forests are made up of trees that shed their leaves during the winter.

29.2    The Sierra Nevada, Rocky Mountains, and Cascades are blanketed by **montane** forests composed of pines, firs, and hemlocks.

29.3    **Trembling aspen** is the most widespread deciduous tree in North America.

29.4    South of Alaska, **high** precipitation, **high** humidity, and **warm** temperatures result in a plant community called the **temperate rain forest**.

29.5    If a pine forest has a well developed canopy it lacks **lower strata** of vegetation.

29.6    In spruce forests, as you descend through the canopy, the amount of sunlight **decreases**/increases.

29.7    In a spruce forest temperature in the upper canopy is **greater** than the temperature in the lower canopy.

29.8    In some coniferous forests colonies of **lichen**, living on leaves in the canopy, fix atmospheric **nitrogen**.

29.9    The debris left on the ground after logging encourages **fires**.

29.10   The prevention of light ground fires in some fir and spruce forests has led to an **increase** in plant diseases and insects, especially the **spruce budworm**.

29.11   In Europe, the deciduous forest ecosystems are represented by two major forest types, the **beech-oak-hornbeam** and the **oak-hornbeam**.

29.12   The eastern slopes of the Sierra Nevada are too dry to support a **montane** coniferous forest, but a **temperate woodland** grows there.

29.13   An uneven-aged deciduous forest usually has **four** strata, the **upper canopy**, the **lower canopy**, the **shrub layer** and the **ground** layer.

29.14    Studies on the nature of nutrient cycling in conifer forests reveal that conifers are nutrient **accumulators**.

29.15    Much of the deciduous forest of the United States has been cleared for **agriculture**.

## MULTIPLE CHOICE

29.16    Temperate forests include all the following except:
  a)    coniferous forests;
  b)    deciduous forests;
  c)    montane forests;
  **d)    tropical rain forests.**

29.17    The _____, the largest tree of all grows in scattered groves on the west slope of the California Sierra Nevada mountain range.
  a)    sugar pine;
  **b)    sequoia;**
  c)    incense cedar;
  d)    lodgepole pine.

29.18    Heavy rainfall during the winter, dense fog during the summer, warm temperatures, and high humidity describe a:
  a)    montane forest;
  b)    pine forest;
  **c)    temperate rain forest;**
  d)    boreal forest.

29.19    Which of the following is not a class of growth form and growth behavior found in a coniferous forest?
  a)    spire-shaped evergreens;
  b)    deciduous trees with pyramid-shaped open crowns;
  **c)    small deciduous broadleaf trees;**
  d)    trees with straight cylinder-shaped trunks and dense crowns.

29.20    Which statement about the vertical stratification in a coniferous forest is not true?
  a)    vertical stratification is not well developed;
  **b)    in a spruce or fir forest the high crown density leads to well developed lower strata;**
  c)    the litter layer is deep and poorly decomposed;
  d)    pine forests have a dense upper canopy and no lower strata.

29.21　　Which of the following statements is true?
　　　　　　a)　in spruce forests the coolest temperatures
　　　　　　　　are in the lower canopy;
　　　　　　**b)　insect populations in coniferous forests are**
　　　　　　　　**many and often destructive to trees;**
　　　　　　c)　the species diversity of mammals in a
　　　　　　　　coniferous forest is high;
　　　　　　d)　illumination in coniferous forests is most
　　　　　　　　intense during the midsummer, when the sun is
　　　　　　　　directly overhead.

29.22　　Cyanophycophilous lichens:
　　　　　　**a)　fix atmospheric nitrogen;**
　　　　　　b)　are primary food producers;
　　　　　　c)　trap calcium and magnesium;
　　　　　　d)　are microbial decomposers.

29.23　　What human activity has had a significant impact on
　　　　　temperate coniferous forests?
　　　　　　a)　agriculture;
　　　　　　b)　urban sprawl;
　　　　　　c)　development of recreation facilities;
　　　　　　**d)　logging.**

29.24　　You would expect to find a riparian woodland:
　　　　　　a)　in semi-arid desert regions;
　　　　　　b)　on the unglaciated Appalachian plateau;
　　　　　　**c)　along the banks of rivers and streams;**
　　　　　　d)　at the point where a boreal forest meets a
　　　　　　　　deciduous forest.

29.25　　The most important mechanism for cycling nitrogen from
　　　　　vegetation to soil in a broad-leaf forest is:
　　　　　　a)　leaching of nutrients from the leaves;
　　　　　　**b)　the death of the small lateral roots of**
　　　　　　　　**broad-leaf trees;**
　　　　　　c)　decomposition of litterfall;
　　　　　　d)　uptake from mineral soil.

29.26　　Which of the following is not a broad-leaf forest?
　　　　　　a)　temperate woodland;
　　　　　　b)　temperate evergreen forest;
　　　　　　c)　Eastern beech and hickory forest;
　　　　　　**d)　redwood forest.**

29.27　　The eastern slopes of The Sierra Nevada support:
　　　　　　**a)　temperate woodlands;**
　　　　　　b)　temperate evergreen forests;
　　　　　　c)　oak-hickory forests;
　　　　　　d)　magnolia-oak forests.

29.28    Which statement about the structure of a broad-leaf
         forest is false?
         a)    the highest temperatures are in the upper
               canopy;
         b)    the lowest humidity levels are near the
               forest floor;
         c)    old-growth, uneven-aged deciduous forests
               consist of four strata;
         **d)    the greatest concentration of life in the
               forest is found in the upper canopy.**

29.29    Which of the following elements could become critically
         low in a northern hardwood forest?
         a)    potassium;
         b)    calcium;
         c)    magnesium;
         **d)    b and c, but not a;**
         e)    a, b, and c.

29.30    Deciduous forests have been highly modified by which of
         the following human activities?
         a)    logging;
         b)    forest fires;
         c)    land clearing;
         d)    introduced diseases and insects;
         **e)    all of the above.**

**TRUE/FALSE**

29.31    T  **F**    All temperate forests are characterized by very
                  little aboveground biomass.

29.32    **T**  F    Montane forests in the Sierra Nevada and the
                  Cascades are composed of hemlock, red fir, and
                  lodgepole pine.

29.33    T  **F**    The vertical stratification in temperate forests
                  is very well developed.

29.34    **T**  F    The litter layer in coniferous forests is thick
                  and decomposition is slow.

29.35    T  **F**    The amount of light reaching the ground in a
                  coniferous forest fluctuates seasonally.

29.36    **T**  F    Mites are many in the litter in a coniferous
                  forest, but earthworms are few.

29.37    T  **F**    Coniferous forests support many and diverse
                  mammalian populations.

29.38　**T**　F　Dead trees and snags form a critical resource in a coniferous forest ecosystem because they provide nesting and den sites for many birds and mammals.

29.39　**T**　F　Montane coniferous forests and pinelands are important sources of lumber.

29.40　T　**F**　Foresters are careful to replant harvested land with a complex array of different tree species similar to the types that were removed.

29.41　T　**F**　Sound conservation practices in the last decade have saved much of the broad-leaf forests in the United States.

29.42　T　**F**　Ecologists have collected very little data on the energy flow and nutrient cycling in deciduous forests.

29.43　**T**　F　In deciduous forests the litter layer is the most important short-term source of nutrients.

29.44　T　**F**　The most important mechanism for cycling nitrogen from vegetation to soil in a deciduous forest is the leaching of nutrients from leaves.

29.45　**T**　F　The species composition of most North American hardwood forests has been greatly altered by lumbering practices and fires.

## MATCHING

**Answer these questions about an old-growth coniferous forest and a new-growth coniferous forest.**

A.　old-growth forest
B.　new-growth forest

29.46　__A__　largest total biomass.

29.47　__B__　highest percentage of living biomass.

29.48　__A__　greatest amount of nitrogen is in the foliage.

29.49　__A__　returns more nitrogen to the forest floor.

29.50　__B__　lowest volume of woody debris and wood litterfall.

## DISCUSSION

29.51　Describe the similarities and differences in nutrient cycling in a broad-leaf forest and a coniferous forest.

29.52    What human activities have had the greatest impact on
         the coniferous forest of the United States?  Has the
         human impact been the same in broad-leaf forests?
         Explain your answer.

29.53    How does the structural stratification of a coniferous
         forest differ from that of the stratification of a
         deciduous forest?

29.54    How do old growth coniferous forests differ from new-
         growth coniferous forests?

29.55    How is the diversity of animal life in a broad-leaf
         forest influenced by the stratification and the growth
         forms of the plants?

# Chapter 30

## Tropical Forests

<u>SENTENCE COMPLETION</u>

30.1    The model of a tropical forest is a **tropical rain forest**.

30.2    In a tropical rain forest there is little annual variation in **temperature** and **rainfall**, but there are great variations in **heat, precipitation**, and **humidity**.

30.3    The largest tropical rain forest in the world is the **Amazon** forest of **South America**.

30.4    **Gallery forests**, like riparian woodlands are found along waterways.

30.5    The **multi-layered lowland rain forest** is a subtype of a rain forest characterized by having luxuriant plant growth.

30.6    Both the **semi-evergreen** and **semi-deciduous** forests are subject to periodic droughts for two to three months a year.

30.7    A tropical rain forest can be divided into **five** general layers, called **strata**.

30.8    Many tropical rain forest plants have elongated, downward curving leaves called **drip tips**. The shape of these leaves helps reduce nutrient **leaching** from the soil.

30.9    **Lianas** include climbers like vines that cling to tree trunks to elevate themselves into the forest canopy.

30.10   Orchids are **macroepiphytes** because they root themselves in niches on the trunks, limbs, and branches of trees.

30.11   Orchids take nutrients, water, and some photosynthate from their host tree. Therefore, they are **hemiparasites**.

30.12   Many tropical forest plants produce prop roots called **buttresses** to give them support in soil that offers poor root anchorage.

30.13   As the lateral buds of a young forest tree begins to grow the tree becomes **sympodial**, the growth pattern typical of mature trees.

30.14    Young tropical forest trees are **monopodial**, that is they consist of a single stem and a tall narrow crown.

30.15    Within the forest crown there is very little leaf and branch overlap because of a growth pattern called **growth shyness.** This growth pattern reduces the competition for **sunlight.**

30.16    The animals feeding above the tropical forest's canopy include **insectivores** and **carnivorous birds.**

30.17    Ninety percent of all the world's **primates** live in tropical rain forests.

30.18    The mean annual net production of a tropical forest is **greater** than that of a boreal forest.

30.19    There are many examples of mutualistic interactions in tropical rain forests. Over 98% of all the flowering plants in the forest are pollinated by **animals.**

30.20    Many people living in tropical rain forests practice a type of agriculture called **slash-and-burn.**

## MULTIPLE CHOICE

30.21    Which of the following abiotic factors is not important when deciding the distribution of tropical forests?
      a)    temperature;
      b)    rainfall;
      **c)    humidity;**
      d)    all of the above.

30.22    Tropical rain forests consist of three groups. The largest tropical rain forest is found in:
      a)    Asia;
      **b)    South America;**
      c)    China;
      d)    Africa.

30.23    Gallery forests are found:
      a)    at high altitudes;
      **b)    along waterways;**
      c)    bordering deserts;
      d)    on leeward sides of mountains.

30.24    In a tropical rain forest the uppermost layer consists of:
      **a)    tall trees over 80 meters high;**
      b)    shrubs and ferns;
      c)    trees with conical crowns;
      d)    plants with downward curved leaves.

30.25    Lianas are:
        a)    stranglers;
        b)    epiphytes;
        **c)    climbers;**
        d)    hemiparasites.

30.26    Stranglers start life as:
        **a)    epiphytes;**
        b)    parasites;
        c)    climbers.

30.27    The soil in tropical rain forests is:
        a)    saline;
        b)    very fertile;
        c)    composed of several layers of rich humus;
        **d)    infertile with most of the nutrients concentrated in the upper 0.3 meters.**

30.28    As you move through the layers of a tropical rain forest beginning at ground level, the level of carbon dioxide and humidity _____ and temperature and evaporation_____.
        **a)    increases, decreases;**
        b)    increases, increases;
        c)    decreases, decreases;
        d)    decreases, increases.

30.29    Which animal type would be most numerous in a tropical ran forest?
        a)    primates;
        b)    reptiles;
        **c)    insects;**
        d)    birds.

30.30    Flowers heavily scented with the odor of carrion are probably pollinated by:
        a)    bats;
        **b)    beetles;**
        c)    birds;
        d)    bees.

## TRUE/FALSE

30.31  T  **F**   Tropical rain forests are restricted to a climatic zone between 30 degrees N and 10 degrees S latitudes.

30.32  T  **F**   Tropical rain forests form a continuous belt of vegetation around Earth's equatorial region.

30.33 **T** F   Tropical forests are found around the terrestrial equator region of Earth, except in areas that experience long annual droughts.

30.34 **T** F   The tropical rain forests in the Indo-Malaysian area has the greatest diversity of plant species.

30.35 **T** F   Seasonal rain forests are subject to droughts every two to four months and are composed of trees that shed their leaves during the dry season.

30.36 T **F**   Dry tropical forests comprise the major plant community of the Amazon Basin and the southern tip of India.

30.37 **T** F   The tropical rain forest consists of at least five distinct layers of vegetation.

30.38 T **F**   Many tropical rain forest plants are stranglers that all begin their life as climbers.

30.39 **T** F   Epiphytes are plants with aerial roots that use crevices on trees as a place to grow.

30.40 **T** F   A tropical forest is one of the most productive ecosystems in the world.

30.41 T **F**   The soil of tropical rain forests is very fertile and suitable for agriculture.

30.42 T **F**   The standing crop biomass of a temperate hardwood forest is twice that of a tropical rain forest.

30.43 **T** F   Flowers pollinated by birds produce no scent but are colored red, orange, or yellow.

30.44 T **F**   Clearing the vegetation in a tropical rain forest could provide good agricultural land for some emerging Third World nations.

30.45 **T** F   Nutrient recycling in a tropical rain forest is dependent on mycorrhizal fungi that transfer nutrients directly from dead organic matter to the roots of the plant.

## MATCHING

For the following questions consider the ecology of the layers in a tropical rain forest.

    A.    uppermost layer
    B.    second tree layer
    C.    lowest tree stratum
    D.    shrubs, tall herbs, fern layer
    E.    ground layer

30.46  __D__   many plants with long downward pointed leaves.

30.47  __A__   conditions here are similar to open land.

30.48  __E__   two to three percent of the incident radiation reaches this level.

30.49  __C__   the deepest layer, well defined and made up of trees with conical crowns.

30.50  __B__   consists of mop-crowned trees less than 50 meters tall.

## DISCUSSION

30.51   There are many different types of tropical forests. Describe rain forests, seasonal forests, and dry forests. How are they similar and how are they different?

30.52   List several plant and animal adaptations that help organisms adapt to the conditions of a tropical rain forest.

30.53   "Tropical rain forests are the most productive ecosystems on Earth." What evidence can you provide to support this statement?

30.54   How do tropical rain forests maintain their nutrient balance? What special role do mycorrhizal fungi play in the nutrient balance?

30.55   What may be some global environmental consequences of destroying the world's tropical rain forests?

# Chapter 31

## Lakes and Ponds

<u>SENTENCE COMPLETION</u>

31.1     **Turbidity** refers to the cloudiness of water and it is influenced by the amount of **silt** suspended in the water.

31.2     Water reaches its maximum density at **four degrees Celsius.**

31.3     The **epilimnion** is a layer of lighter, warm water near the surface of a lake.

31.4     For every **one** meter downward the temperature of a lake declines **one** degree Celsius.

31.5     The **hypolimnion** is a layer of cold water at the bottom of a lake.

31.6     The **fall overturn** takes place when surface water cools to a point where temperature is uniform throughout the lake.

31.7     Rooted plants around the edge of a lake comprise the **littoral** zone.

31.8     **Nekton** are free-swimming organisms such as fish.

31.9     The **benthic** zone comprises the bottom of a lake.

31.10    **Fish** make up the largest group of nekton in the limnetic zone.

31.11    Life in the **profundal** zone is only abundant during the spring and fall overturn.

31.12    The dominant organisms in the benthic zone are **anaerobic bacteria.**

31.13    **Periphyton** include algae and diatoms that are attached to submerged rocks and sticks.

31.14    The **detrital food chain**, not the grazing food chain dominates lakes and ponds.

31.15    Lakes that are rich in nutrients are called **eutrophic** systems.  Lakes which are nutrient-poor are **oligotrophic.**

31.16    A eutrophic lake:
        a)    has a low surface-to-volume ratio;
        b)    is deficient in nutrients;
        c)    is free of phytoplankton;
        **d)    there is no correct answer.**

31.17    Dystrophic systems:
        a)    have brown-colored water stained by humic material;
        b)    have highly productive littoral zones;
        c)    have a low planktonic production;
        d)    a and b, but not c;
        **e)    a, b, and c;**

31.18    In most oligotrophic lakes _____ is a limiting nutrient.
        **a)    phosphorus;**
        b)    nitrogen;
        c)    calcium;
        d)    carbon.

31.19    The primary production in the littoral zone of a lake is provided by:
        a)    phytoplankton;
        **b)    macrophytes;**
        c)    nekton;
        d)    aufwuchs.

31.20    In lake ecosystems detrital metabolism takes place in the:
        a)    littoral zone;
        b)    epilimnion;
        **c)    benthic zone;**
        d)    thermocline.

31.21    Phytoplankton in a lake are fed upon by:
        a)    vertebrate planktivores;
        b)    periphyton;
        **c)    zooplankton;**
        d)    invertebrate planktivores.

31.22    The benthic zone:
        a)    is rich in aerobic bacteria;
        **b)    harbors communities of periphytons;**
        c)    is usually higher in oxygen than the littoral zone;
        d)    is rich in phytoplankton.

31.23    Which of the following would be classified as nekton?
         a)   water insects;
         **b)   lake trout;**
         c)   polychete worms;
         d)   anaerobic bacteria.

31.24    The _____ contributes heavily to the input of
         organic matter into a lake.
         a)   limnetic zone;
         b)   profundal zone;
         **c)   littoral zone;**
         d)   thermocline.

31.25    Life is abundant in the profundal zone:
         a)   during the winter months;
         b)   during the summer;
         **c)   during the spring and fall overturn;**
         d)   all year.

**TRUE/FALSE**

31.26  T  **F**   The water at the bottom of a lake is called the
                 thermocline.

31.27  **T**  F   Water in a lake reaches its maximum density at
                 four degrees Celsius.

31.28  **T**  F   Just below the epilimnion is a layer of heavy cool
                 water called the metalimnion.

31.29  T  **F**   During the summer the oxygen level in the water of
                 a lake is highest at the bottom.

31.30  **T**  F   The shallow water surrounding a pond or lake
                 comprises the littoral zone.

31.31  T  **F**   Photosynthesis is greater than respiration at the
                 compensation level of light.

31.32  **T**  F   Fish in the littoral zone lack strong lateral
                 muscles.

31.33  T  **F**   The profundal zone is the major contributor of
                 organic matter to a lake.

31.34  **T**  F   The algae and diatoms at the bottom of a lake are
                 part of the periphyton.

31.35  T  **F**   Lakes are totally self-contained ecosystems and
                 are not influenced by the terrestrial ecosystems
                 that surround them.

31.36  **T**  F    Lakes, like terrestrial communities, are dominated by detrital food chains.

31.37  T  **F**    The macrophytes provide the primary production for the limnetic zone of a lake.

31.38  T  **F**    Macrophytes are affected by the concentrations of dissolved nutrients in the open water of a lake; it is from that area that they draw nitrogen and phosphorus.

31.39  **T**  F    Piscivores feed on vertebrate planktivores.

31.40  **T**  F    Eutrophic lakes are shallow and warm and have very little oxygen at their bottoms.

## MATCHING

A.   littoral zone
B.   limnetic zone
C.   profundal zone
D.   benthic zone

31.41  __A__    fish in this zone have compressed bodies so they can squeeze through the thick vegetation.

31.42  __E__    the dominant organism found here are anaerobic bacteria.

31.43  __B__    the main forms of life in this zone are phytoplankton and zooplankton.

31.44  __C__    this is the zone beyond the depth of effective light penetration.

31.45  __C__    its energy source is a rain of organic material produced in the layer above.

## DISCUSSION

31.36    Describe the biological importance of upwelling and overturn in a lake.  How will overturn influence productivity?

31.37    How is the turbidity of water related to the compensation depth (level)?

31.48     Compare and contrast an oligotrophic lake to a
          eutrophic lake in relation to:

          a.   abundance of nutrients and plankton.
          b.   surface-to-volume ratio.
          c.   abundance of fish.
          d.   size of the littoral zone.
          e.   type of bottom consistency in the benthic
               zone.

31.49     Describe the vertical zonation of a lake or pond.  How
          does the penetration of light and the rate of
          photosynthesis influence the characteristics of the
          vertical zones?

31.50     Nutrients move between terrestrial ecosystems and
          aquatic ecosystems through different pathways.
          Describe examples of each of the these nutrient
          pathways:

          a.   biological
          b.   geological
          c.   meteorological
          d.   hydrological.

# Chapter 32

## Freshwater Wetlands

### SENTENCE COMPLETION

32.1    Wetlands support **hydrophytic** vegetation that includes pondweeds and cattails.

32.2    **Obligate** wetland plants include submerged pondweeds, bulrushes, and the bald cypress.

32.3    **Vegetation, hydrologic conditions,** and **soil properties** are all important factors to consider when defining a wetland.

32.4    Wetlands that develop in shallow upland depressions are called **basin wetlands.**

32.5    **Riverine** wetlands develop along periodically flooded river banks.

32.6    A **quaking bog** has a floating mat of peat over its open water.

32.7    Hydroperiod included **duration, frequency, depth,** and **season of flooding.**

32.8    **Zonation** reflects the response of plants to hydroperiod.

32.9    The **muskrat** is a prairie marsh herbivore that has been introduced into Eurasia.

32.10   Some wetland plant species have a **small** root biomass in the summer because nutrients are transported to the **aboveground** biomass.

32.11   Nitrogen moves between a wetland and the atmosphere by **nitrogen fixation, denitrification,** and **nitrification.**

32.12   Peat bogs depend on **precipitation** to provide the plants with nutrients.

32.13   The primary production in peatlands is **low.**

32.14   Wetlands act as **water-filtration systems** because they take up excessive nutrients and heavy metals incorporating them into plant tissue and deposit them in the bottom mud.

32.15     **Agriculture** has led to the drainage of many hectares of wetlands.

## MULTIPLE CHOICE

32.16     Hydroperiod includes which of the following?
          **a)     frequency of flooding;**
          b)     water chemistry;
          c)     direction of water flow;
          d)     kinetic energy of water.

32.17     You would find sphagnum moss growing in:
          a)     a mangrove swamp;
          b)     a fresh water marsh;
          c)     a saline flat;
          **d)     a bog.**

32.18     Forested wetlands are called:
          a)     marshes;
          b)     mires;
          **c)     swamps;**
          d)     fens.

32.19     Hydrophytic plants include:
          a)     cacti;
          b)     chaparral plants;
          **c)     cattails;**
          d)     grasses.

32.20     Peatlands:
          a)     have a rate of organic production that
                 exceeds the rate of decomposition;
          b)     may consist of acid-forming plants;
          c)     are the home for several insectivorous
                 plants;
          **d)     all of the above.**

32.21     _____ make up a large component of the animal
          life in a wetland.
          a)     mammals;
          b)     carnivores;
          **c)     herbivores;**
          d)     scavengers.

32.22     Wetlands have been drained and the land used for:
          a)     agricultural land;
          b)     solid waste dumps;
          c)     housing developments;
          **d)     all of the above.**

32.23    Wetlands once made up _____ of Earth's surface.
        a)    10%;
        **b)    3%;**
        c)    25%;
        d)    53%.

32.24    Wetlands can be used as sites for growing:
        **a)    cranberries;**
        b)    raspberries;
        c)    blackberries;
        d)    thimbleberries.

32.25    Which of the following is a mammalian inhabitant of wetlands?
        a)    moose;
        b)    muskrat;
        c)    hippopotamus;
        d)    a and b, but not c;
        **e)    a, b, and c.**

## TRUE/FALSE

32.26    T  **F**    Hydrophytic vegetation will not grow in water-saturated soil.

32.27    T  **F**    Sedges and alders are obligate wetland plants.

32.28    **T**  F    The red maple is a facultative wetland plant, growing both in dry uplands and forested wetlands.

32.29    T  **F**    Basin wetlands develop along shallow flooded river banks.

32.30    **T**  F    Marshes are composed of emergent herbaceous vegetation.

32.31    **T**  F    Riparian woodlands grow along large river systems.

32.32    **T**  F    A moor is a blanket bog.

32.33    T  **F**    The hydroperiod of fringe wetlands does not undergo seasonal fluctuations.

32.34    T  **F**    Peat bogs are abundant in tropical regions.

32.35    T  **F**    Wetlands are biological deserts because they support very few plant and animal populations.

32.36    **T**  F    Wetlands are sedimentary or detrital systems.

32.37  T  **F**   The average maximum standing crop of biomass in a wetland is lower than the annual aboveground productivity.

32.38  **T**  F   Carnivorous plants use insects as a source of nitrogen.

32.39  T  **F**   The primary production of a peatland is high.

32.40  **T**  F   Wetlands can filter heavy metals from water and reduce the pH of industrial wastewater.

## MATCHING

A.  shrub swamp
B.  saline flats
C.  swamp
D.  bogs
E.  fen

32.41  __D__   waterlogged soil with a spongy covering of <u>Sphagnum</u> moss.

32.42  __A__   waterlogged soil with alder, willow, and dogwoods.

32.43  __E__   a mire fed by water moving through mineral soil, dominated by sedges.

32.44  __B__   inland saline area that floods after a rain. Common plants include salt grass, and saltbush.

32.45  __C__   a forested wetland.

## DISCUSSION

32.46   From a hydrological point of view there are three major types of wetlands.  What are they?

32.47   How does the physical aspect of water and hydroperiod influence the structure of a wetland?

32.48   What is a peatland?  How do peatlands differ from other freshwater wetlands?

32.49   Wetlands support a rich biological community of plants and animals.  Why is this so?  What group of animals make up the most conspicuous component of animal life in a wetland?  Why?

32.50   How are phosphorus and nitrogen cycled in a wetland?

# Chapter 33

## Flowing-Water Ecosystems

**SENTENCE COMPLETION**

33.1      The land that a river drains is its **watershed**.

33.2      The character and the structure of a river is determined by the **velocity** of the current.

33.3      Flowing-water ecosystems consist of two habitats, the **turbulent riffle** and the **quiet pool**.

33.4      **Riffles** are the sites of primary production in a stream.

33.5      The **periphyton** of a stream occupies the same position as the phytoplankton of a lake.

33.6      The pools of a stream are the sites of **decomposition**.

33.7      Most of the carbon dioxide in a stream occurs as **carbonate** and **bicarbonate salts**.

33.8      Fish inhabiting fast-water streams have strong **lateral** muscles to keep them in place in the water.

33.9      Much of the energy source in the lotic system is **detrital material**.

33.10      **Fungi** are important in the breakdown of CPOM because the large particles offer sufficient surface area for mycelium to develop.

33.11      Craneflies, caddisflies, and stoneflies are member of a large feeding group, the **shredders**, which feed on coarse particulate matter.

33.12      Some **filtering collectors**, such as the caddisfly, spin a net to capture small organic particles floating in the water.

33.13      The algal coating on stones in a stream are fed upon by the **grazers**.

33.14      The **composition of the bottom** and the **width** of a stream are two variables that influence its production.

33.15      Headwater streams are **heterotrophic** and are dependent on the input of **detritus** from terrestrial vegetation along the stream.

33.16    The velocity of a stream is influenced by:
         a)    the width of the stream;
         b)    the depth of the stream;
         c)    the composition of the bottom;
         d)    a and b, but not c;
         **e)    a, b, and c.**

33.17    The primary production of a stream takes place:
         a)    on the shore;
         b)    in ponds;
         **c)    in the riffles;**
         d)    in the watershed.

33.18    If the water of a stream has a high pH the richer it is
         in:
         a)    oxygen;
         **b)    carbonates;**
         c)    nitrogen;
         d)    phosphorus.

33.19    The fine particulate organic matter (FPOM) is broken
         down and recycled by:
         **a)    bacteria;**
         b)    fungi;
         c)    shredders;
         d)    grazers.

33.20    Pools with sandy bottoms:
         a)    offer substrate for the attachment of
               aufwuchs;
         b)    are more productive than pools with gravel
               bottoms;
         c)    offer protected places for insect larvae to
               hide;
         **d)    are not productive because they do not
               provide a substrate for organisms to attach
               to.**

33.21    Headwater streams:
         a)    are heterotrophic;
         b)    are heavily dependent on detritus from
               vegetation along its banks;
         c)    have few grazers, but many shredders;
         **d)    all of the above.**

33.22    As streams increase in width to medium-sized rivers:
         a)    water temperature decreases;
         **b)    primary production is dependent on algae and
                rooted aquatic plants;**
         c)    shredders are the dominant consumers;
         d)    predators are cold-water species.

33.23    If the water used as a coolant in a power plant is
         released into a river, it:
         a)    will pose little threat to aquatic life;
         b)    will raise the oxygen content of the water;
         **c)    will lower the dissolved oxygen of the water;**
         d)    it will increase the pH of the water.

33.24    Dams:
         a)    reduce the flow of sediments reaching
               riparian habitats;
         b)    alter the invertebrate and vertebrate
               populations of streams;
         c)    decrease the volume of water flow in the
               stream;
         **d)    all of the above.**

33.25    Most of the energy available for stream food chains
         comes from:
         a)    primary production by algae in the water;
         b)    photosynthesis by aquatic bacteria;
         **c    terrestrial sources;**
         a)    the rapid breakdown of detritus by grazers
               and gougers.

## TRUE/FALSE

33.26  **T**  F   The velocity of the water in a stream is the most
                  important factor deciding the structure of the
                  waterway.

33.27  T  **F**   Most of the primary production of a stream takes
                  place in quiet pools of water along the route of
                  the stream.

33.28  **T**  F   During the fall and summer the pools along a
                  stream are the major sites of carbon dioxide
                  production.

33.29  T  **F**   The churning and swirling of stream water over
                  riffles causes a significant drop in the amount of
                  dissolved oxygen in the water.

33.30  T  **F**   A lotic system is largely autotrophic.

33.31 **T** F     Energy in a lotic system is lost through geological and biological pathways.

33.32 **T** F     Shredders are organisms that feed on coarse particulate organic matter (CPOM).

33.33 **T** F     Shredders pass off over 50 percent of the material they ingest as feces.

33.34 T **F**     Invertebrates that burrow into water-logged limbs and trees are called gougers.

33.35 **T** F     The CPOM, FPOM, and invertebrates tend to float downstream thus creating a sort of traveling benthos.

33.36 T **F**     Streams with sandy bottoms are more productive than streams with gravel bottoms.

33.37 **T** F     As a stream increases in width to a medium-sized river, the riparian vegetation becomes the major energy and nutrient contributor.

33.38 T **F**     Near the mouth of a river the current becomes slower, sediments accumulate on the bottom, and the water system becomes heterotrophic.

33.39 **T** F     Warm water released into a cool-water ecosystem can be considered a type of water pollution.

33.40 **T** F     The Colorado River is the most regulated river in the world.

## MATCHING

**A major energy source for a lotic system is detrital material carried to it from adjacent ecosystems.**

        A.   CPOM
        B.   FPOM
        C.   DOM

33.41   **B**   the most important nutrient source at the mouth of a river, used by bottom-dwelling collectors.

33.42   **C**   a major source is from rainwater that drips from leaves overhanging a stream and carries nutrients dissolved in water droplets.

33.43   **B**   bacteria feed on this detrital material.

33.44   **A**   fed upon by shredders.

33.45 __A__ represented by leaves and twigs dropped into the water by vegetation growing along the stream's bank.

## DISCUSSION

33.46 What are some major problems that organisms must face living in a moving-water environment? What are some special adaptations found in animals in this environment? Name some types of aquatic life, tell where in the stream they are found, and describe their adaptations for stream living.

33.47 How do seasonal changes influence the movement of energy though food webs in lotic systems?

33.48 How do lotic ecosystems keep nutrients upstream? How do the mechanisms reduce nutrient flow downstream? Compare a lotic nutrient cycle with a lentic nutrient cycle.

33.49 What is spiraling? Explain the pathway that a nutrient or atom would take in a spiral. Describe the spiraling of phosphorus as analyzed at Oak Ridge, Tennessee.

33.50 "The lotic ecosystem is a continuum of changing environmental conditions from its headwaters to its mouth." Would you agree with this statement? Describe how a stream's physical and biological character changes along its length.

# Chapter 34

## Oceans

**SENTENCE COMPLETION**

34.1    **Sodium** and **chlorine** make up about 86 percent of sea salt.

34.2    The amount of chlorine in sea water is used as an **index of salinity**.

34.3    When surface water is replaced by water moving upward from the deep, the process is known as **upwelling**.

34.4    When Earth, moon, and sun are in line with each other, the gravitational pull of the sun and moon result in very high tides known as **spring tides**. When the gravitational pull of the sun and moon interferes with each other a **neap tide** results.

34.5    As sea water cools it becomes **heavier**.

34.6    The open body of water that makes up the ocean is called the **pelagic zone**, The bottom region of the ocean is the **benthic zone**.

34.7    A subdivision of the pelagic zone that runs from the surface to about 200 meters is the **photic zone**.

34.8    The major herbivores in the pelagic zone are **zooplankton**.

34.9    During the summer **dinoflagellates** may become toxic to marine organisms.

34.10   **Nekton** are organisms that are free-swimming.

34.11   In the **bathypelagic** regions bioluminescence is used to attract prey, escape predators, and recognize individuals.

34.12   The **epifauna** and **epiflora** include animals and plants that line the rocky bottoms of the ocean.

34.13   **Infauna** burrow into the substrate at the bottom of oceans.

34.14   **Bacteria** in the benthic zone can synthesize protein from dissolved nutrients and thus become important food for deposit feeders.

34.15    The most productive regions of the ocean are shallow
         **coastal waters,** but the productivity of the open ocean
         is **low.**

## MULTIPLE CHOICE

34.16    The most abundant element in sea water is:
         **a)     chlorine;**
         b)     potassium;
         c)     calcium;
         d)     sodium.

34.17    Salinity of the ocean is affected by:
         a)     evaporation;
         b)     movement of water masses;
         c)     precipitation;
         d)     the mixing of water masses of different
                salinity;
         **e)     all of the above.**

34.18    Primary productivity of the ocean is carried out by:
         a)     dinoflagellates;
         b)     zooplankton;
         **c)     phytoplankton;**
         d)     epifauna.

34.19    Mixed tides are common in:
         a)     the Gulf of Mexico;
         b)     the Atlantic Ocean;
         **c)     the Pacific Ocean;**
         d)     the Arctic Ocean.

34.20    You would expect the number of phytoplankton to be most
         numerous in the _____ zone.
         a)     bathypelagic zone;
         b)     mesopelagic zone;
         **c)     photic zone;**
         d)     benthic zone.

34.21    Organisms that are bioluminescent live in the:
         **a)     bathypelagic zone;**
         b)     mesopelagic zone;
         c)     photic zone;
         d)     pelagic zone.

34.22    Most sea plants are small because:
         a)    they cannot photosynthesize easily in salt
               water;
         b)    salt water is deficient in nutrients;
         c)    **small organisms can absorb more nutrients
               than larger ones;**
         d)    the surface area of their bodies is small and
               they have trouble absorbing solar energy.

34.23    Which of the following is not an example of
         zooplankton?
         a)    copepods;
         b)    gastropods;
         c)    **dinoflagellates;**
         d)    comb jellies.

34.24    Nekton include which of the following?
         a)    sharks;
         b)    baleen whales;
         c)    penguins;
         d)    **all of the above;**
         e)    none of the above.

34.25    The largest part of the energy flow in pelagic
         ecosystems depends on:
         a)    copepods;
         b)    zooplankton;
         c)    **bacteria and protists;**
         d)    dinoflagellates.

**TRUE/FALSE**

34.26  T  **F**   The index of salinity of water is determined by
                  measuring the concentration of sodium.

34.27 T  **F**   Seawater becomes lighter as it cools and reaches
                  its freezing point a 0 degrees Celsius.

34.28  **T**  F   Pressure changes in the ocean are much greater
                  than on land.

34.29  T  **F**   The oceans of Earth experience one high tide and
                  two low tides every 24 hours.

34.30  T  **F**   Neap tides are most likely to occur when the moon
                  is full.

34.31  **T**  F   Water that overlies the continental shelf is part
                  of the neritic zone.

34.32  T  **F**   The mesopelagic zone is an area that shows sharp
                  seasonal variation in temperature and light.

34.33　T　F　Zooplankton are the major herbivores in a pelagic ecosystem.

34.34　T　F　The distribution and composition of phytoplankton is influenced by light, temperature, and nutrients.

34.35　T　F　Eupausiids, copepods, and planktonic arthropods represent the zooplankton in the ocean.

34.36　T　**F**　Burrowing worms and clams living the mud in the floor of the ocean are part of the epifauna.

34.37　T　F　Sediments below 6000 meters in the ocean contain little organic matter, but do contain large quantities of silica and aluminum oxides.

34.38　T　F　Near hydrothermal vents in the ocean chemosynthetic bacteria act as primary producers oxidizing reduced sulfur compounds.

34.39　T　F　The giant clam, Calyptogena magnifica has a symbiotic relationship with a chemosynthetic bacteria.

34.40　T　**F**　Tropical seas have a thermal stratification that varies with the seasons and as a result such waters experience frequent upwellings.

**MATCHING**

      A.　photic zone
      B.　mesopelagic zone
      C.　bathypelagic zone
      D.　benthic zone
      E.　neritic zone

34.41　__D__　the ocean bottom.

34.42　__E__　water over continental shelves.

34.43　__C__　complete darkness is the order here, except for some bioluminescence.

34.44　__A__　water near the surface to 200 meters with sharp gradients on light intensity, temperature, and salinity.

34.45　__B__　contains the oxygen-minimum layer and high concentrations of phosphate and nitrate.

## DISCUSSION

34.46 It was surprising for biologists to find that there was a great diversity of life living in deep ocean trenches. What reasons have been given to explain this diversity?

34.47 Why is it that most fisheries are located within 200 miles of the coast?

34.48 Why do tides occur at different times each day?

34.49 Why are pelagic plants small? Why do they live in the photic zone? What is the ecological importance of the compensation depth?

34.50 What reproductive strategies would you expect to find among animals living in the bottom of a deep sea?

# Chapter 35

## Intertidal Zones and Coral Reefs

<u>**SENTENCE COMPLETION**</u>

35.1    The first major zone we see as we make the transition
        from land to sea is called the **supralittoral fringe**.

35.2    Below the black zone lies the **littoral zone**, an area
        covered and uncovered by the daily tides.

35.3    **Kelp** is a large brown algae found in the infralittoral
        zone.

35.4    <u>Gigartina</u> is a **red** algae often found growing in
        association with **mussels** in the littoral zone.

35.5    The productivity of a kelp bed is **greater**/less than the
        productivity of a tropical rain forest.

35.6    The nature of a beach is influenced by the size of the
        **sand particles**.

35.7    **Epifauna** and **infauna** make up the two important groups
        of life on sandy and muddy beaches.

35.8    The **meiofauna** live within the sand of a beach and range
        in size from 0.5 mm. to 62  m. long.

35.9    A sandy beach can be divided into **supralittoral**,
        **littoral**, and **infralittoral zones**.

35.10   As mud and sand lose oxygen from the respiration of
        bacteria and chemical oxidation a layer of **iron
        sulfides** forms.

35.11   The energy base for a sandy beach is organic matter
        that is first attacked by **bacteria**.

35.12   **Deposit feeders** obtain food by burrowing through sand,
        ingesting substrate, and removing organic matter.

35.13   Coral reefs are produced when dead skeletal material of
        **carbonate-secreting organisms** accumulates on stable
        geologic foundations.

35.14   **Barrier** reefs form parallel to shorelines and are
        separated from land by lagoons.

35.15   Corals are both **photosynthetic** and **heterotrophic**
        organisms.

35.16   Zooxanthelleae are:
        a)   symbiotic;
        b)   can photosynthesize;
        c)   endozoic dinoflagellates;
        **d)   all of the above;**
        e)   none of the above.

35.17   Coral reefs are:
        a)   unstable ecosystems that will soon give way to climax ocean communities;
        b)   not highly productive ecosystems;
        **c)   highly productive ecosystems;**
        d)   unable to retain nutrients within the system and continuously lose nutrients to the surrounding seas.

35.18   _____ is the most serious intertidal pollutant.
        a)   methyl mercury;
        b)   DDT;
        **c)   oil;**
        d)   sewage.

35.19   The black zone:
        a)   is created by oil that has washed onto beaches;
        **b)   is the home of lichens and blue-green algae;**
        c)   is covered and uncovered by changes in the tide;
        d)   is the habitat favored by oysters and mussels.

35.20   Dense mats of _____ provides homes for echinoderms in the lower reaches of the littoral zone.
        **a)   mussels;**
        b)   kelp;
        c)   rockweeds;
        d)   <u>Laminaria</u>.

35.21   Tidepools are subject to wide and sudden fluctuations in:
        a)   temperature;
        b)   salinity;
        c)   pH;
        d)   oxygen levels;
        **e)   all of the above.**

35.22    A sandy beach can support:
        a)    large populations of algae;
        b)    crabs;
        **c)    infauna occupying tubes and burrows;**
        d)    coralline red algae.

35.23    The _____ zone of a beach is the zone where true marine organisms first appear.
        a)    littoral;
        **b)    supralittoral;**
        c)    black;
        d)    infralittoral.

35.24    _____ are horseshoe-shaped rings of coral reefs surrounding a lagoon formed by a volcano.
        a)    barrier reefs;
        b)    fringing reefs;
        **c)    atolls.**

35.25    _____ provide(s) an important energy allotment to intertidal life.
        a)    sunlight;
        b)    chemosynthetic bacteria;
        **c)    waves;**
        d)    algae.

**TRUE/FALSE**

35.26   T  **F**    The greatest number of coral species is found at the crest near the surface of a reef.

35.27   **T**  F    Grazing by sea urchins and some fish will encourage the growth of encrusting coralline algae in coral reefs.

35.28   **T**  F    Fringing coral reefs grow out from the rocky shores of islands.

35.29   **T**  F    Periwinkles are highly resistant to desiccation.

35.30   T  **F**    If the wave action along a rocky shore is severe you will find large populations of periwinkles and little algal growth.

35.31   T  **F**    As the carbon dioxide level of a tidepool increases at night the pH of the water rises.

35.32   **T**  F    Wave energy on a rocky shore is approximately twice that of solar radiation.

35.33   T  **F**    The highest standing crop of kelp is found in areas receiving little or no wave action.

35.34  **T**  F    Interstitial fauna is characterized by elongated bodies, reduced setae, spines and tubercles and no pelagic larval form.

35.35  **T**  F    You would expect to find beach hoppers and ghost crabs in the supralittoral zone of a beach.

35.36  T  **F**    Organisms living within the sand or mud of a beach must tolerate violent fluctuations in temperature and salinity.

35.37  **T**  F    There is an inverse relationship existing between the turbulence of the water and the amount of organic matter that accumulates on a beach.

35.38  **T**  F    Sandy beaches are essentially heterotrophic and are very dependent on organic matter that has been produced away from the beach.

35.39  T  **F**    You would find zooanthellae living in the gastrointestinal tract of periwinkles.

35.40  **T**  F    The supralittoral fringe is the first major zone you see when you move from land to sea.

## MATCHING

**The rocks along the shore are exposed by ebbing tides. Answer the following questions about these zones.**

    A.   supralittoral zone
    B.   littoral zone
    C.   infralittoral zone

35.41  __C__   home of kelp and bladder rockweed.

35.42  __A__   often called the spray zone because water rarely covers this area except during high spring tides.

35.43  __B__   area covered and uncovered by daily tides, the upper regions of this zone would be dominated by periwinkles.

35.44  __C__   under water most of the time, becoming exposed only at extreme low tide. Area subject to violent wave and current action.

35.45  __B__   area is dominated by barnacles.

**DISCUSSION**

35.46   What role do waves play in the ecology of a rocky shoreline?

35.47   How do wind, waves, and tides affect the plants and animals living in the intertidal zone?

35.48   The primary productivity of a coral reef is high. Why is this the case?

35.49   What are the environmental factors that are necessary for coral reefs to develop? What are the types of coral reefs and under what conditions will they develop?

35.50   What type of human activities offer a threat to the survival of intertidal zones and coral reefs?

# Chapter 36

## Estuaries, Salt Marshes, and Mangrove Forests

### SENTENCE COMPLETION

36.1    **Estuaries** are ecosystems where fresh water joins salt water.

36.2    **Deltas** form at the mouth of rivers as stream-carried sediments are deposited in quiet waters.

36.3    Most of the organisms living in an estuary are **benthic.**

36.4    Estuaries have both **plankton** and **detrital** food chains.

36.5    The producers in estuaries are **dinoflagellates** and **diatoms.**

36.6    **Tides** and **salinity** are the two important abiotic factors shaping the character of salt marshes.

36.7    The zonation of salt marsh plants is strongly influenced by **tides.**

36.8    The **fiddler crab** is an omnivorous feeder that is one of the most adaptable animals living in a salt marsh.

36.9    The salt marsh is a **detrital** system, most of the organic material coming from <u>**Spartina**</u>.

36.10   Up to **47** percent of the net primary production of a salt marsh is respired by **microorganisms.**

36.11   Coastal marshes are the major wintering grounds for **waterfowl.**

36.12   In tropical regions tidal flats are occupied by **mangrove forests.**

36.13   Many species of trees growing in mangals have root extensions called **pneumatophores** that can take in oxygen.

36.14   **Mangals** form where wave action is lacking and sediments accumulate.

36.15   In an estuary, at the mouth of a river, fresh water from the river is **less**/more dense than salt water and thus it sits **above**/below the sea water.

36.16    In estuaries, tidal overmixing takes place:
- **a)** **during a flood tide;**
- b) when fresh water sinks below a wedge of seawater;
- c) when the salinity of estuary waters changes during the fall and spring;
- d) when evaporation of fresh water from the estuary exceeds the inflow of fresh water from the river discharge.

36.17    Estuarine organisms must solve, which of the following problems?
- a) how to maintain a constant body temperature;
- b) how to adjust to changing salinity;
- c) how to prevent being carried away by the currents;
- d) a, b, and c;
- **e)** **b and c, but not a.**

36.18    Some estuary fish, such as the striped bass spawn:
- a) in saltwater;
- b) in freshwater;
- **c)** **at the point where freshwater meets saltwater;**
- d) in lakes.

36.19    One of the largest invertebrate communities in an estuary is made up of:
- a) mussels;
- **b)** **oysters;**
- c) clams;
- d) periwinkles.

36.20    The major producer component of an estuary consists of:
- a) kelp;
- b) blue-green algae;
- **c)** **dinoflagellates and diatoms;**
- d) rockweeds.

36.21    In the Florida keys the _____ dominates the estuarine world.
- a) seagrass community;
- b) salt marsh community;
- c) mudflat community;
- **d)** **mangrove community.**

36.22    A  flat, grass-covered coastal area with estuarine
         ecosystems in temperate climates is called a:
         **a)**    **salt marsh community;**
         b)    mudflat community;
         c)    mangrove community;
         d)    salt barren community.

36.23    <u>Spartina</u> <u>alterniflora</u>:
         **a)**    **can tolerate being inundated by salt water**
                   **many hours each day;**
         b)    can tolerate being flooded only for a few
               hours each day;
         c)    can tolerate being flooded only at extreme
               high tides, usually once a month;
         d)    cannot tolerate any flooding and must grow in
               soil free of any water.

36.24    Cord grass can thrive in salty environments that are
         alternately flooded and exposed by tides because:
         a)    they can secrete salt from their leaves;
         b)    their stems contain air tubes that can carry
               oxygen to their roots;
         c)    they have developed a mutualistic
               relationship with nitrogen-fixing bacteria in
               the mud;
         d)    a, b, and c;
         **e)**    **a and b, but not c.**

36.25    Which of the following adaptations would you expect to
         find in plants living in a mangrove forest?
         a)    salt pores located on the leaves to secrete
               excess salt;
         b)    waxy leaves to reduce water loss;
         c)    seeds specialized to germinate while attached
               to the parent tree;
         d)    roots adapted to anchor the plant in soft
               mud;
         **e)**    **all of the above.**

<u>**TRUE/FALSE**</u>

36.26   **T**  F    Deltas develop at the mouths of rivers.

36.27   T  **F**    Salinity in an estuary remains stable and
                    constant, showing no seasonal or daily
                    fluctuations.

36.28   **T**  F    Nutrients and oxygen are carried into an estuary
                    with incoming tides.

36.29   **T**   F    The decay of <u>Spartina</u> contributes to the detritus in coastal waters, which serves as an important food for many marine organisms.

36.30   T   **F**    In estuaries planktonic organisms attach themselves securely to the mud on the bottom.

36.31   T   F    Organisms living in estuaries are essentially marine and must be able to tolerate seawater.

36.32   T   **F**    Anadomous fish spend their entire life cycle in estuaries.

36.33   T   F    Diatoms manufacture high caloric fats and lipids rather than low-energy carbohydrates.

36.34   T   F    Salt marsh vegetation exists in distinct zones, determined by the plants' tolerance to being covered by water.

36.35   T   F    The salt marsh is one of the most productive habitats in the marine environment.

36.36   T   **F**    In salt marshes nitrogen fixation offsets denitrification, so marshes do not require nitrogen input into the system.

36.37   T   F    The marsh animals that are most adaptable have both lungs and gills so they can extract oxygen from water and from air.

36.38   T   F    The highest rate of productivity in mangrove forests occurs in those forests that experience daily tides.

36.39   T   F    The most massive destruction of mangals occurred when herbicides where used during the Vietnam War.

36.40   T   F    Mangrove trees have leaves equipped with salt pores that help maintain a proper salt balance in the plants.

<u>MATCHING</u>

     A.   low marsh zone
     B.   high marsh zone
     C.   tidal creeks
     D.   salt meadow
     E.   salt pans

36.41 __A__ anaerobic mud, marsh periwinkle, _Spartina alterniflora._

36.42 __C__ mud algae, diatoms, dinoflagellates, outflow of fresh water.

36.43 __B__ floods a few hours each day, short form of _Spartina alterniflora_, low tidal exchange, short, open canopy.

36.44 __D__ _Spartina patens_, mounds of detritus, mice and insects are common visitors.

36.45 __E__ algal crust, crystallized salt, edges support _Salicornia_ and _Distichlis_.

## DISCUSSION

36.46 What role does detritus play in the food webs of estuaries?

36.47 How would the productivity of an estuary be affected if tidal lands were drained for urban development?

36.48 What are some special adaptations of _Spartina alterniflora_ that allows it to adjust to the abiotic conditions of a salt marsh?

36.49 What are some adaptations mangrove trees have that allow them to survive in a tropical wetland?

36.50 Describe the zonation in an estuary, salt marsh, and mangal. What abiotic factors are most important in deciding the zonation in each of these ecosystems?